Praise for the previous editions of *Light Years*

'A fascinating book on a fascinating subject. It brings together all aspects of light in an unusual and compelling way.'
Sir Patrick Moore

'Light's properties often seem mysterious to the point of being unfathomable. Yet in this extraordinary book Brian Clegg manages to explain them through the lives of those so fixated with light that they have shaped our perception of it … Clegg's accessible writing style manages to encapsulate the lives of light's disciples with humorous and interesting anecdotes … [He] also provides real scientific insight into how light behaves. He explains complex theories through lucid metaphors, without resorting to the elaborate diagrams so beloved of some popular science writers … Clegg indulges in future gazing, too, the results of which are quite awesome …'
Karen Peploe, *New Scientist*

'This immensely likeable work of pop science traces "man's enduring fascination with light", from Aristotle's plans for a death ray (burning enemy ships with a giant array of mirrors) through to a recent experiment that seems to have sent Mozart's 40th Symphony faster than light, and thus back through time. Clegg is very good at explaining the bizarre properties of light …' Steven Poole, *The Guardian*

'A fascinating, non-technical treatment of the concept of light … an excellent resource … makes for compelling reading.' *ScienceScope*, the magazine of the US National Science Teachers' Association

LIGHT YEARS

LIGHT YEARS

THE EXTRAORDINARY STORY OF MANKIND'S FASCINATION WITH LIGHT

BRIAN CLEGG

ICON

This revised edition published in the UK in 2015 by
Icon Books Ltd, Omnibus Business Centre,
39–41 North Road, London N7 9DP
email: info@iconbooks.com
www.iconbooks.com

Originally published in 2001 by Piatkus,
and in a fully revised version in 2008 by Macmillan

Sold in the UK, Europe and Asia
by Faber & Faber Ltd
Bloomsbury House, 74–77 Great Russell Street,
London WC1B 3DA or their agents

Distributed in the UK, Europe and Asia
by TBS Ltd, TBS Distribution Centre, Colchester Road,
Frating Green, Colchester CO7 7DW

Distributed in the USA by
Consortium Book Sales & Distribution
34 13th Avenue NE, Suite 101, Minneapolis, MN 55413

Distributed in Australia and New Zealand
by Allen & Unwin Pty Ltd, PO Box 8500,
83 Alexander Street, Crows Nest, NSW 2065

Distributed in South Africa by
Jonathan Ball, Office B4, The District,
41 Sir Lowry Road, Woodstock 7925

Distributed in Canada by Publishers Group Canada,
76 Stafford Street, Unit 300
Toronto, Ontario M6J 2S1

ISBN: 978-184831-814-4

Typeset in Janson Text by Marie Doherty

Printed and bound in the UK
by Clays Ltd, St Ives plc

Contents

About the author

Science writer Brian Clegg studied physics at Cambridge University and specialises in making the strangest aspects of the universe – from infinity to time travel and quantum theory – accessible to the general reader. He is editor of www.popularscience.co.uk and a Fellow of the Royal Society of Arts. His previous books include *Inflight Science*, *Build Your Own Time Machine*, *The Universe Inside You*, *Dice World*, *The Quantum Age*, *Science for Life* and *Introducing Infinity: A Graphic Guide*.

www.brianclegg.net

Preface

And God said
'Let there be light':
And there was light.

GENESIS 1:3

L ight is something that we take for granted. It is a fact of
life, available at the press of a switch. It is the absence of
darkness, the everyday gift of the Sun. It is a small part of the
physics we are taught at school, a thing of ray diagrams and
geometry, a natural phenomenon without substance. But light
is not so easily compartmentalized. Its beguiling combination
of fragility and endurance, of delicacy and power, captures the
imagination just as it has fascinated scientists through the ages.

For thousands of years, uncovering the nature of light has
proved an irresistible challenge. It forms a scientific quest that
has endured from the conjectures of the ancient Greeks to the
work of twentieth century geniuses like Albert Einstein and
Richard Feynman. By combining light's history with the latest
research we can assemble a complete picture of this remarkable
phenomenon and its place at the centre of creation.

What first was seen as merely the mechanism of sight has
proved to be so much more. The source of all life on Earth,
providing warmth, powering the weather, driving the photosynthetic process that generates oxygen. The self-sustaining
interplay of magnetism and electricity that lies behind Einstein's

special relativity. The fundamental glue that keeps all matter together. And perhaps even the key to time itself.

Looking back at the life and work of the extraordinary people who have uncovered light's secrets provides both an understanding of light and a front row seat in the development of the remarkable new light-based technologies that are appearing as we enter the twenty-first century. Technologies that have the potential to transform reality itself.

When I was at university studying physics, I was overwhelmed by the power and beauty of light, yet so much that I read at the time on this remarkable subject made it seem dull. You only have to look at optical diagrams with rays and lenses and focal points to feel a yawn coming on. Coming back to light now has been wonderful, a chance to rekindle the amazement and delight I felt 30 years ago. That sense of wonder is what *Light Years* is all about.

Acknowledgements

Thanks to all those who have helped in the production of various editions of this book, including my former agent Peter Cox, Sara Abdulla and Duncan Heath.

Specific thanks to Professor Edward H. Adelson, for permission to reproduce his stunning optical illusion, and Professor Günter Nimtz for his considerable input and helpful comments on the manuscript. And a final thank you to the many individuals who have patiently helped with information and assistance. It would be boring to list them, but they know who they are.

Chapter One

At the speed of light

For now we see through a glass, darkly.
St Paul

Imagine this. The dawn light is creeping into your room. You get up from your bed and open the curtains. Outside the window, the inferno of an active volcano distorts the air. A river of red-hot lava is streaming down the scarred mountainside. A rain of ash falls near the window, yet you hear nothing, feel nothing.

Quickly, you move to the second window and pull back the curtain. Here, even though it's morning, the sky is black, a crisper black than you have ever seen. The stars stand out, laser sharp. Before you is a rugged, near-white plain, surrounded by impossibly high, needle peaks. And then your eye is caught by something else. Standing out from the blackness is a bright circle of blues and greens with traceries of white. You are seeing the Earth from the surface of the Moon.

Nervously, half-expecting the air to rush out of the room, you open the window, to be struck by a burst of vertigo. Behind the glass is a leaden yellow-grey sky, hanging heavy over the already-bustling city streets 25 floors below. Nothing that you saw through the window glass exists. There is no volcano, no lunar landscape; there are no stars.

A magical tunnel

Close the window again and still the Earth is riding serenely
in the sky. It's as if the window's glass were not looking out of
the side of the building, but opening instead onto a magical
tunnel leading straight onto the surface of the Moon. There
are no video screens or electronics involved, just glass with
very special properties. This is slow glass, first dreamed up by
the 1970s visionary writer Bob Shaw. A special glass that light
takes months or even years to pass through.

With such remarkable glass it would only take a good site
in front of a beautiful view to create such stunning windows. If
light takes a year to get from one side of the glass to the other,
then one year after it is put in position, the first glimpse of the
landscape will reach the other side. As the light takes a year to
pass through, everything that has been happening in front of
the glass will be seen a year later behind it. Shift the glass into
a building and it carries a year's worth of light with it. You've
got a window on an exotic location for as long as it takes the
remaining light to make its slow journey through the material.

The ultimate speed

It was only in the late 1990s that technology caught up with
the imagination and made slow glass a possibility. The discoveries described in this chapter demonstrate the remarkable
power of the new light technology. Later, we will plunge back
2,500 years to follow the story of humanity's fascination with
light. In that story, light's immense speed will be a recurring
theme. For slow glass it presents a particular problem.

A beam of light travels at around 300,000 kilometres each

second in the vacuum of space, a speed that belies our experience of nature. A hummingbird's wings flap 4,200 times in a minute, near invisible to the human eye. Yet in the duration of a single flap of those wings, a beam of light could have crossed the Atlantic Ocean. On 20 July 1969, Apollo 11 landed on the Moon after a journey of four days. If it had, instead, set off for the nearest star, Alpha Centauri, which light takes four years to reach, the Apollo capsule would still be travelling after a thousand millennia.

In glass, light moves a little slower than it does in space, but it would still require a window 5,000,000,000,000 kilometres thick to hold a year's worth of light. If slow glass is to be made, there is an enormous challenge to face. There has to be a way to apply the brakes, to slow light down by a factor of a billion billion or more. Unlikely though this sounds, in the late 1990s a substance was created that can do just that.

Einstein's strange matter

The substance with the amazing effect on light is a strange form of matter called a Bose–Einstein condensate (physicists have to have a particularly good day to come up with a snappy name like 'photon' or 'quark'). We are used to matter coming in three types – solid, liquid and gas. Since the 1920s it has been known that there is a fourth form of matter, generated in the raging nuclear furnace of the Sun – plasma. This is the next stage beyond a gas, where the easily removed electrons have been broken off the atoms and the result is a soup of ions – atoms with some electrons missing – and the electrons themselves.

The four states of matter – solid, liquid, gas and plasma – have a startling parallel in a theory developed over 2,000 years ago. The Greek philosopher Empedocles thought that every-thing was made up of four elements – earth, water, air and fire – each equivalent to one of the modern states. Some of the ancients thought there should be a fifth element, the substance from which the heavens were constructed, called the quintes-sence. This handily corresponds with a then-hypothetical fifth state of matter that Einstein dreamed up. The idea also dates back to the 1920s, when a young Indian physicist called Satyendra Bose wrote to the world-famous scientist describ-ing his ideas. Einstein would have received many letters from scientific hopefuls, but this one caught his attention. Bose had found a totally new way to describe light.

Thanks to Einstein's theories, light had begun to be thought of as photons – tiny, insubstantial particles that shot through space like bullets from a gun. Bose experimented with describ-ing light mathematically as if these photons were a collection of particles that was already well understood – a gas. Einstein helped Bose firm up the maths, but was also inspired to imagine a fifth state of matter. By applying intense cold or pressure to a material he believed that it would eventually reach a state where it would no longer be an ordinary substance; instead it would share some of the characteristics of light itself. Such a state of matter is a Bose–Einstein condensate, the material that could provide the key to producing slow glass.

Nearly 80 years after the theory was developed, a Danish scientist has used a Bose–Einstein condensate to drag the speed of light back to a crawl. Her name is Lene Vestergaard Hau, one of the few women to take an active part in the history of light. In 1998, Hau's team set up an experiment where two lasers were blasted through the centre of a vessel containing sodium atoms that had been cooled to form a Bose–Einstein condensate.

Normally the condensate would be totally opaque, but the first laser creates a sort of ladder through the condensate that the second light beam can claw its way along – at vastly reduced speeds. Initially light was measured travelling at around 17 metres per second – 20 million times slower than normal. Within a year, Hau and her team, working at Edwin Land's Rowland Institute for Science at Harvard University, had pushed the speed down to below a metre per second – and more was to follow, as we will discover later.

Hau's material is not quite slow glass. There is one more problem to overcome. Imagine you had a piece of special glass one centimetre thick that took a year for light to get through. It would live up to expectations if you were looking straight through the glass. But things would be different when looking at the edges of the scene. Now the light is arriving at an angle, passing through more of the glass before it gets to you. It could easily travel through half as much glass again, and so take half as long again to get through. With an ordinary window the difference isn't noticeable, but light that hits slow glass at an angle would take months longer to arrive than the light that arrived straight on. Views from every direction would appear at different times, combining images to produce a nightmare confusion.

To overcome this effect, a slow glass window has to do more than just let light through. It needs to capture the whole image at the surface of the window, whatever the angle the light has arrived from. That total view then must pass through the window as a piece, rather than as masses of uncoordinated rays heading in all directions. This requirement isn't as impossible as it sounds. It is very similar to the way in which a hologram is produced, combining the rays of light from different directions to make a unified picture. In the hologram, this gives a three-dimensional view that changes as you move, built into a flat, two-dimensional photograph. It is such an image, with three dimensions compressed into two, that would have to be sent through the window. The combination of holographic techniques and a very slow material would deliver true slow glass.

While the technology required at the moment to have such control on light's speed is formidable, the mere fact of its existence gives some hope that in the future slow glass may move from fiction to practical reality. The first lasers, after all, were heavy-duty, complex devices requiring conditions that were inconceivable outside the laboratory – yet some modern lasers can fit on a pinhead and are happy to function in the unprotected environment of a consumer product like a laser pointer.

Breaking the light barrier

If the possibilities of slow glass, bringing light to a virtual standstill, are fascinating, then taking the opposite tack, pushing light above its normal speed, has even more remarkable consequences. As we will explore in detail in Chapter 8, Einstein's special theory of relativity showed that light was the fastest thing in existence. Nothing, he argued, could exceed that 300,000 kilometres per second. According to the special theory of relativity, any solid object approaching the speed of light would get heavier and heavier until its mass was infinite. Even the speed of an insubstantial snippet of information should never get past the 300,000 kilometres per second barrier, because the peculiarities of relativity mean that a faster than light signal would travel backwards in time. If it *were* possible to broadcast a message fast enough, we could use light to say hello to our ancestors.

Such technology would transform human existence. If a signal could be sent back even a tiny fraction of a second it would make it possible to build computers that worked thousands of times faster than current machines, limited as they are by the speed of internal communication. With information sent even further back, disasters could be averted by broadcasting warnings. All gambling based on prediction, from the roulette wheel to the stock market, would be destroyed. There is hardly an aspect of life that would not be fundamentally changed. Yet this is not the most dramatic implication of sending a message back in time.

The very foundations of reality would come under threat. Being able to send a message into the past would shatter the rigid connection of cause and effect. For most scientists this is enough to prove that getting a message past light speed is impossible. It's not that they have any objection to getting a sneak preview of lottery results this way, but rather that bewildering paradoxes emerge when information is sent backwards through time.

Time COPs

It is easy to feel the impact of the paradox by considering a simple time transmitter that could send a radio message back just a few seconds. This transmitter is fitted with a radio control, so it can be switched on and off remotely. At noon precisely, the transmitter is used to send a message back in time. This message is the signal to the transmitter's own radio control. When the message is received at five seconds before noon, it switches the transmitter off. Now, when noon arrives, the transmitter is switched off. So how could the message have been sent?

But if the message wasn't sent, the transmitter would still be switched on.

Rather than deal with such mind-bending possibilities, physicists resort to the 'Causal Ordering Postulate', sometimes known as the time COP. This sounds impressive, but amounts to little more than saying that effect never can come before cause. (It's actually a little more sophisticated, allowing the effect to precede the cause if there's no way the effect can influence the cause, but the result is the same.) It follows that anything that would endanger the relationship of cause and effect, like sending a message back in time, is impossible. Professor Raymond Chiao of the University of California, a leading exponent of superluminal physics – the science of faster than light motion – believes there is no way to send a message back through time. But Chiao's own experiments in the late 1990s opened a loophole in the light speed barrier.

At the sub-microscopic level of photons, the minuscule particles that make up a beam of light, the everyday expectations of the world fall apart. The familiar, predictable behaviour of objects disappear, leaving only probability and uncertainty. This is the world of quantum physics, discovered by Max Planck and Albert Einstein around a hundred years ago. Thanks to the bizarre nature of reality at the quantum level, individual photons of light have a small but real chance of jumping through solid objects and appearing on the other side in a process known as tunnelling.

Quantum short cut

Tunnelling emerges from the bizarre statistical view that quantum mechanics takes. Generally speaking, quantum mechanics expects, just as we would in the normal world, that when a car drives into a wall it bounces back. Every now and then, though, quantum theory says it should pass straight through. The probability is incredibly low – far less than winning a lottery week after week after week – but it exists. In a beam of light there are many, many photons, and the chance that a single photon will cross an apparently impenetrable barrier is much higher than that of a whole car jumping through a wall. This phenomenon, tunnelling, has been widely observed. In fact, if it weren't for tunnelling, there would be no life on Earth.

The light of the Sun that heats the Earth and triggers the release of oxygen through photosynthesis is produced by a deceptively simple process. In the intense furnace of the core of a star (like the Sun), charged particles of the most basic element, hydrogen, combine to make helium, the next element up the chain. In this process energy is released. The reaction can only happen if hydrogen particles come into close contact, but each particle is positively charged. These charges repel each other, like magnets when the same poles are brought together. Even in the Sun's heart, the particles can't combine – just as well, or there would be an immense explosion, burning out the Sun in a second as all the hydrogen was converted. The repelling force forms a barrier that has to be overcome to form helium, just as we have to fight against gravity to jump over a physical barrier like a fence. It is the strange reality of quantum physics that makes this possible. Some hydrogen particles jump through the barrier to fuse together – they have tunnelled.

To give an accurate picture of what is happening, we should really drop the term 'tunnelling'. It implies slowly grinding your way through an obstacle. What really happens is much more startling. At one moment a particle is one side of the barrier, the next it is on the other. It jumps rather than tunnels, but instead of flying over a physical barrier, it actually passes from one position to the other without moving through the points in between. This instant jump means that any photons travelling along a path that includes a barrier to tunnel through manage to get along that path at faster than the speed of light.

Chiao and his team demonstrated this strange phenomenon, measuring light travelling at 1.7 times the normal speed. If this light beam could be made to carry a signal, that message would, according to relativity, be shifted backwards in time. But Professor Chiao wasn't worried about destroying the fabric of reality. His experiment relied on generating individual photons, and the mechanism that made this possible provided no way of controlling when a photon would emerge. Without such control, the photons could not carry a message. Equally, there was no way of deciding which photons would get through the barrier – most don't – and so it seemed impossible to keep a signal flowing. Without the ability to send a message there would be no chance of disrupting causality.

At the time, Professor Chiao was unaware of developments in another laboratory in Cologne, Germany. The refined tones of Mozart's 40th Symphony, clearly a message, were about to be transmitted at four times the speed of light. The stakes for reality were about to be raised.

But before exploring the nature of these faster than light experiments and how they could pose a threat to existence itself, we need to do some time travelling of our own, taking a 2500 year trip back to a time when the very existence of light seemed as close to magic as it did to science.

Chapter Two

The philosophers

The Atoms of Democritus
And Newton's Particles of light
Are sands upon the Red Sea shore,
Where Israel's tents do shine so bright.
WILLIAM BLAKE

In Neolithic Britain, from around 3000 BC, Stonehenge acted as both a temple of light and a marker of the important seasonal changes predicted by the movement of the Sun. By the time Stonehenge was at its peak, 1500 years later, the Egyptians linked the Sun with a god – *the* god, Ra, creator of the Universe, first among the deities. The Sun was thought to be Ra's eye, the source of all life and creation. Light and warmth poured from the god, at the same time a gift and something to fear. In a papyrus dating from 1300 BC, a priest-scribe noted the thoughts of Ra himself:

I am the one who opens his eyes and there is light.
When his eyes close, darkness falls.

To the Egyptian people, always poised on the balance between flood and drought, Ra was generous, but also terrible. To look directly into the eye of the god was to be blinded. Even to glance at his glory caused pain. When the ancient Egyptians dabbled with monotheism it was the Sun's disk, the Aten, that was the focus of their worship. The temples at Akhetaten, the newly built capital dedicated to the Sun, carried a eulogy to the benefits of Aten:

> You are beautiful, great, shining and high above every land, and your rays enfold the lands to the limit of all you have made … You, sole god, who no other is like.

Though not directly a part of God, light was still given great significance in the early Jewish beliefs that gave rise to modern Judaism, Christianity and Islam. In the dramatic biblical Genesis creation story, light is a part of the first day of time, along with the Earth, the heaven and the waters. Later, as the ancient Greek civilization formed the foundations of modern Western culture, light reappeared in the religion and legends of the Greeks.

Flying too close to the Sun

The picture most of us have of ancient Greek religion is a fuzzy mix of childhood stories, more like a work of fiction than a religious text. It's easy to think of this impression as a reflection of our ignorance, flavoured by Disney cartoon characters and Hollywood epics. But that picture is surprisingly accurate. There was no written core of the Greek religion – no equivalent of the Bible or the Qur'an – instead, there was a complex web of myth: stories told to illustrate the nature of divinity, always combining entertainment with education. This flexible structure meant that even the gods themselves changed with time. Originally it was Helios who rode the Sun's chariot across the sky, but he became absorbed into the central figure of Apollo, son of Zeus.

In these ever-changing and developing tales, the most striking human interaction with light came in the experience of Daedalus and Icarus. Daedalus was an inventor, said to have designed the labyrinth for King Minos of Crete that contained the half-man, half-bull monstrosity, the Minotaur. Architects of secret structures of the time were sometimes killed to destroy their knowledge. It might have been better for Minos if Daedalus had suffered that fate. The inventor gave the secret of the labyrinth to the king's daughter, Ariadne, who passed it on to her lover Theseus. After Theseus managed to kill the Minotaur and escape, Daedalus was imprisoned along with his son Icarus.

Daedalus built wings of wax and feathers so the two of them could fly away to safety, but Icarus was too bold, revelling in the freedom of flight as he soared higher and higher. Forgetting his purpose, he trespassed on the territory of the Sun. The heat of the Sun's fierce rays melted the wax on his wings, leaving Icarus to plunge to his death in the sea. This story was a pointed parable, demonstrating the dangers of the quest for knowledge and of wanting too much for the self. It has been used countless times since to provide imagery to illustrate this risk. Yet it was not long after the myth probably first emerged, around the seventh century BC, that the Greeks began a dogged pursuit of knowledge.

The development of philosophical thought as a legitimate activity was triggered by a change in circumstances of the Greek people. From around 650 BC, the aristocratic groups that had been in control were overthrown by tyrants. Given the connotations this word has today, it's perhaps surprising that these tyrants were largely welcomed – the 'tyrant' label only meant that they had seized power unofficially, not that their actions were oppressive. Likely to be wealthy commoners, the tyrants proved popular, supporting trade and encouraging the economy. With new political and trading strength came the opportunity to take a step back, to take time to think, rather than being concerned with mere survival. This new ease of living made it possible for the Greeks, always a people inclined to structure, to develop the schools of philosophy.

Light – in fact, all of nature – came to be treated in a new way. There was still religious awe, but alongside it was room for practical curiosity and logic. The religion did not go away (though not every philosopher subscribed to religious beliefs), but now there was something more. By 500 BC, light was being considered in some detail, particularly by Empedocles. By a coincidence, this interest was echoed on the far side of the Earth. The followers of Chinese philosopher Mo-Tzu had also taken on the challenge of light.

Light from the east

Unlike the relative calm of the Greeks, China was in upheaval. Though the great Zhou dynasty that had seen the first unification of that huge country had more than 200 years still to run, Chinese society was breaking into factions. The political instability of the time seemed to demand a very practical philosophy. It saw the rise of Legalism, a school that prided itself on efficient, soulless statecraft rather than purity of thought. Yet it also produced Mo-Tzu, who has been described as China's first true philosopher in the Western sense.

Mo-Tzu was said to be a disgruntled follower of Confucius,

who became frustrated by his master's aristocratic leanings and emphasis on ritual. Mo-Tzu's philosophy combined pragmatism and an emphasis on universal love. Those who followed Mo-Tzu took an approach to light that was every bit as practical as Legalism's transformation of bureaucracy into a high art. They measured and observed, noting how flat and curved mirrors produced different types of reflection. They found that by letting light shine through a tiny pinhole in a piece of wood they could project a weak, upside-down image onto a white surface. This discovery was the earliest known camera obscura, a device that remained popular into Victorian times, and eventually gave rise to all our present day photographic equipment.

The inner light

By contrast with his Chinese counterparts, Empedocles did not experiment; to do so was alien to Greek values. Instead he looked inside himself for his inspiration. Light and sight seemed inextricably entwined, so Empedocles pictured light as a beam of fire transmitted from the eyes. The two approaches, Chinese and Greek, were the total opposite of the modern stereotype that labels Eastern culture as inward looking and contemplative while the West is considered obsessed with externals, with measurement and analysis.

This tendency to ignore experiment was not quite as irrational as it seems to modern eyes. The Greeks argued that our senses were easy to fool. What we experienced was not always a helpful guide. It was more important to look within. It's certainly true that the senses have their limits. Optical illusions like the one in Figure 2.1, produced by Professor Edward H. Adelson at MIT, demonstrate just how fallible our sense are.

Figure 2.1 Checkerboard optical illusion
(courtesy of Professor H. Adelson)

In this picture, the squares labelled A and B are exactly the same shade of grey. Because of the way our brain is programmed to handle objects and shadows we are fooled into thinking that the square labelled B is much lighter, but it isn't. (If you don't believe this is true, see the animation at http://www.universeinsideyou.com/experiment3.html which moves square A alongside square B. They really are the same shade.) Unfortunately, the limitations of our senses don't make our mental processes any more accurate, but the Greeks believed that the only hope of finding the truth was through pure reason.

Empedocles, born around 492 BC, had a privileged upbringing in Acragas (now Agrigento) on the Sicilian coast. His rich family was prepared to indulge the passionate enthusiasms

that soon brought him to the attention of others. His followers considered him a seer, but according to historian George
Sarton, contemporary critics, perhaps detecting a tendency in
Empedocles to make sure that the dice rolled his way, considered him a charlatan. Certainly Empedocles had a high opinion
of his own worth. In later years he flaunted many of the trap-
pings of royalty, from a purple robe and golden girdle to a
constantly present group of fawning attendants.

There is little doubt, though, that Empedocles was driven
by fervent curiosity and could not have been satisfied with the
easy life of home. He travelled the Greek world in a search
for knowledge. He was an archetypal educated man of his culture, embodying a fascination with the nature of the world.
But for all his enthusiasm and originality, Empedocles brought
the cumbersome baggage of Greek philosophy to his scientific
studies. He had no concept of using experiment to prove ideas
– debate and the application of pure thought were the only
tools he could employ.

Much of Empedocles' time was taken up with medicine. He
seems to have had real skill as a healer, and made the most of
his practical ability to build up his image, passing off cures that
he knew were perfectly natural as miracles. When he wasn't
peddling remedies, Empedocles shared an interest with his
contemporaries in the nature of matter. How was a solid substance made up? Was light itself matter, or something different?
Empedocles' most far-reaching (if wildly wrong) contribution
to this debate was to devise the four elements. Everything, he
decided, could be broken down to the essential components of
earth, air, fire and water. (There's a strong parallel here with
the much earlier Genesis creation story, where the first things
created are the Earth, heavens, light and water.)

There is some logic to Empedocles' theory. For instance, when a piece of wood burned it gave off fire, smoke (a form of air), and ash (a kind of earth). This simplistic picture was taken up by two of the big names to influence Western development – Aristotle and Plato – and became the accepted view for over 2,000 years. It still bizarrely crops up in some New Age and alternative philosophies today.

A poet as well as a philosopher (his songs were the surprise hit of the Olympic Games in 440 BC), Empedocles conjured up a flowery picture of the mechanism of sight. In his book *On Nature* he says that Aphrodite, the goddess of love

> kindle[d] the fire of the eye at the primal hearth of the universe, confining it with tissues in the sphere of the eyeball.

Although this language is poetic, Empedocles was being literal. He envisaged actual fire, passing through special channels to separate it from the waters of the eye and flowing out in a blazing stream to the objects that were seen. Such a dramatic picture has to be understood in the context of his four elements – light had to be composed of fire, as it could hardly be earth, air or water.

This fiery light from the eye, the accepted reality for over a thousand years, seems hopelessly flawed to a modern mind. If light originates in the eyes, why can we not still see when the Sun goes down? Empedocles had not missed the contribution of the Sun. In fact he even suggested that the Earth caused the darkness of night by blocking the Sun's rays, a concept that was well ahead of his time. Yet he was able to separate two quite independent kinds of light in his mind. The sunlight he regarded as a facilitator that enabled the eye's light to make sight possible. Imagine opening a box to let light into it. The action of moving the box lid doesn't generate light, it just makes it possible for light to get in. Similarly, Empedocles believed that the Sun only made it possible for the light from the eye to function correctly.

Empedocles' theory was influenced by more than the practicalities of vision. To Greek thinking, the very nature of what was seen, or at least how it was described, was coloured by this inner view. Homer, writing perhaps 400 years before Empedocles, described the sea as wine-dark – yet no one would now consider the colour of the sea to resemble wine. In fact, both blue and green seem to have represented very different concepts from their present day meanings. The closest word there was in ancient Greek to blue is *kyanos*, which from the context in which it was used suggests darkness rather than a colour.

A similar confusion exists over *chloros*, the word that is most similar to green; it was applied to both blood and honey. It seems that chloros was not really a colour, but rather a state of freshness, of new, growing life. A facile explanation for this attitude to colour would be that the ancient Greeks were more susceptible to colour blindness than we are, but there is no evidence to support this. Instead it seems that the feelings attached to an object were given more significance that any observed colouring. The principal light was the inner light not the outer.

Seeing in the dark

Although Empedocles' theories would not be discarded for a millennium and more, they weren't the only attempt to

describe how light worked. The most significant competition
came from the atomists. This faction was a spin-off from the
school that Pythagoras set up before even Empedocles was
at work. Two fourth century BC philosophers, Leucippus and
Democritus, devised the almost prescient concept that everything was made up of tiny indivisible particles – atoms (literally
'uncuttable' in the Greek original). As they believed that all
creation was constructed in this way, they also thought that
light must consist of such tiny particles, flowing in a stream
like a spray of fine powder from source to observer.

The ideas of the atomists were not forgotten, but always
remained on the fringe of acceptability. Even when Newton
attended Cambridge in the 1600s, the atomist view was not
considered particularly significant, but it appealed very much
to Newton himself, and he was to construct a whole theory of
light that was driven by the atomists' ideals (see Chapter 5). For
the moment, though, it was Empedocles' theory that remained
the accepted truth, reinforced by the contribution of Plato, the
highly influential philosopher born in Athens around 428 BC.

Plato (probably a nickname meaning 'broad shouldered' –
he may actually have been called Aristocles) was the youngest
son of an extremely wealthy family. He dabbled in politics, but
the upheavals following the final Peloponnesian war between
Athens and Sparta made this a dangerous pursuit. The execu-
tion of his philosophical master, Socrates, in 399 BC brought
this message home with terrible force. Socrates was technically
charged with heresy – neglecting the gods and introducing
his own deities – but in reality, his crime was more likely to
have been his active criticism of those in power. Socrates' fate
brought Plato to think that the study of mathematics, science
and philosophy was a safer option.

Although famed as philosopher, Plato's doctrines are not the easiest to pin down as they appear as a series of dialogues – almost as fiction – rather than clear expositions of fact. But some of his scientific views, and specifically that of the mechanics of sight, are more clearly recorded.

Plato was conscious of the problems that the inability to see in the dark presented. He expanded the part of Empedocles' theory that dealt with sight as a special interaction between the light of the eye and the light of the outer world. Plato thought that the two merged into a single link, tying together the object being looked at and the inner person. The conjoining of the two types of light produced, Plato thought, an optical highway to channel information on what was being seen to the soul.

Despite his attempt at rationalization, Plato's view remained purely philosophical. It lacked the mathematical reasoning we would now think of as scientific. But less than a hundred years later, another great Greek name was putting a different spin on the nature of vision. Euclid, working around 300 BC, was two generations on from Plato – in fact he was probably educated by Plato's pupils. That is, if he existed at all.

Euclid's rays

That there can be doubt about the existence of such a well-known historical figure may seem bizarre, but there is insufficient evidence to be sure if the works of Euclid are attributable to a single man, a teacher and his pupils, or even a group of philosophers operating under a fictional name (this has occurred since, when a team of mathematicians published a series of works under the constructed name Bourbaki). This uncertainty makes any biographical information about Euclid at best speculative.

Whatever the reality of his existence, Euclid was obsessed with geometry. He applied the unwavering measure of spatial mathematics to the behaviour of sight. Yet despite this logical approach, Euclid managed to further refine the light-from-the-eyes theory rather than dismiss it entirely.

As Einstein would, more than 2,000 years later, when realizing the unique nature of light's speed, Euclid used a thought experiment, acting out a hypothetical situation in his mind to test his deductions. He imagined looking for a needle that had been dropped on the ground. As he searched, even though he was looking in the right general direction, he didn't see the needle. Then all of a sudden it sprang into view. Euclid reasoned that light from the Sun must always be hitting the needle and reaching the eye, so if that were the *only* light, we ought to be able to see the needle immediately. Sight, though, he argued, was dependent on the sunlight's interaction with a ray that shone from the eye, and that ray needed a conscious focus on the object.

This sounds very similar to Plato's theory, but Euclid's big step forward was the idea that this ray from the eye travelled in a straight line, bounced off the needle (or whatever was being looked at) and was reflected back into the eye. The specifics might have been faulty, but he had painted a picture of light that would make a true scientific view possible. Suddenly light had been transformed from a diffuse vaporous phenomenon to something that travelled in straight lines, its behaviour predictable by the new-fangled mathematics of geometry. That light travelled in straight lines would be a fundamental assumption all the way up to the twentieth century, when the distorting mirror of Einstein's genius would throw even this basic premise into question.

Weapons of light

Shortly after Euclid's time, another great philosopher took the ideas of straight line optics and came close to using them to build a death ray. Born in 287 BC, Archimedes lived practically his entire life in Syracuse in Sicily, though he probably spent some time in Alexandria, as he often exchanged personal letters with mathematicians based there. He is now remembered for his mechanical concepts and for carrying on Euclid's work. Archimedes certainly had an obsessive enthusiasm with geometry. Plutarch, writing 350 years later, wryly observed that Archimedes' servants had to drag him from his work to get him to the baths to wash him, and when he was there, Archimedes would still be drawing diagrams using the embers of the fires, and even marking out lines on his naked body as he was being washed and anointed.

Like Euclid, Archimedes was fascinated by light and particularly by mirrors. He wrote a book on optics, now lost along with all the detail of his optical theories. Archimedes lived in an unsettled time for Greece. The Romans, whom the Greeks had contemptuously dismissed as insignificant barbarians, were sweeping across Greek territories. The once great Hellenic civilization was on the verge of collapse. And Archimedes, for all his genius, ended up in the wrong place at the wrong time. He had designed engines of war that were used to bombard invading ships, but despite these, the Romans seemed unstoppable.

It was 212 BC. With the enemy closing in on Syracuse, Archimedes had the inspiration of using light itself as a weapon. He knew that small, curved mirrors could concentrate the rays of the Sun enough to set kindling alight. This ability to focus energy at a distance seemed an ideal way to attack the Roman's vulnerably flammable wooden ships before they were even in range of his projectile weapons.

Archimedes drew up plans for great curved metal sheets to be fixed in frames on the harbour. These dazzling constructions were intended to capture the Sun's rays, focusing them to a point until the undiluted heat of the day became a miniature furnace. But the mirrors were never made. Perhaps the craftsmen, more used to blacksmithing than precision engineering, found them too much of a challenge. Perhaps the stricken city had lost so much to the war effort that it could not find time and money to construct the mirrors. Perhaps even the great Archimedes was laughed at when he claimed it was possible to destroy the Roman enemies without even touching them.

It may have been the mirrors that Archimedes was still working on in his last minutes. He was said to be drawing and re-drawing diagrams when one of the invading Roman soldiers found him. Without looking up, Archimedes cursed the interruption. 'Do not disturb my diagrams.' They were his last words. The soldier who found the 75-year-old man was in no mood to tolerate such disrespect from a member of a defeated race. Archimedes was slaughtered without compassion.

On the brink

The Romans did not entirely eliminate Greek culture. There remained one last flowering of scientific philosophy before Western civilization took the fall into darkness. The man responsible was Ptolemy, living in the Greek city of Alexandria on the edge of Egypt in the second century AD. It was a time of painful transition. Ptolemy was not Greek in the classical sense.

His name alone suggests it. In fact he is sometimes mistakenly called an Egyptian.

More properly he was Claudius Ptolemaeus, the first name showing his Roman citizenry, the second that he lived in Egypt. Nonetheless he was born in Greece and followed the Greek tradition of scholarship. Hardly anything more is known about Ptolemy as a man, apart from the astoundingly precise dates of his first and last recorded observations at Alexandria on 26 March 127 and 2 February 141. Ptolemy made his name with a major study of astronomy – his system, built on Aristotle's view that the Sun and the planets revolved around the Earth on crystal spheres, would remain the absolute standard for another 1,400 years. But his optical work was equally lasting.

His most significant observation of light was, like his model of the solar system, both influential and wrong. He studied the way a beam of light bent as it passed into water, the process known as refraction. By noticing that a coin seems to move if it is placed at the bottom of an empty cup and then water is poured in, and by adding in Euclid's ideas of straight line optics, he was able to say quite correctly that light bends inwards towards a straight line into the water or glass when it enters a denser material from air. The reverse happens on the way out.

So far, so good. And Ptolemy also listed many measurements that he made in establishing just how much the light was bent – an approach that was totally contrary to the traditional Greek tactic of untested theorising and much closer to the modern scientific method. Unfortunately, his deductions from his data were not correct. He thought that there was a fixed proportion between the angle at which the light hit the material and the angle it bent to when it got inside. While this is almost true for small angles, it gets further and further from reality as the angles get bigger. The whole business of refraction took hundreds of years to untangle.

Ptolemy was not without later detractors. The sixteenth century astronomer Tycho Brahe, who constructed the best maps of the stars to be made before telescopes were available, thought that Ptolemy had passed off measurements of star positions made by the earlier Hipparchus as his own. The data produced by Hipparchus, working around 150 BC, is lost, making it impossible to compare the two. But Ptolemy's observations had a consistent error that suggested he might have copied earlier data, then tried to adjust it for the passage of time. Ptolemy made it clear how much he depended on the work of Hipparchus, leaving himself open to posthumous attack. The ever-volatile Isaac Newton attacked Ptolemy vehemently, accusing him of:

> A crime committed against fellow scientists and scholars, a betrayal of the ethics and integrity of his profession that has forever deprived mankind of fundamental information about an important area of astronomy and history.

Certain that Ptolemy had invented data to fit his theories, Newton went on to say:

> Instead of abandoning the theories, he deliberately fabricated observations from the theories so that he could claim that the observations prove the validity of his theories. In every scientific or scholarly setting known, this practice is called fraud, and it is a crime against science and scholarship.

Modern scholars are less critical, accepting that Ptolemy added valuable observations to knowledge that was already in the public domain. As for Ptolemy's book on optics, no one in the modern world has ever seen it, so it may seem odd that we can be sure that he listed detailed experimental data. Most of the copies of his book were destroyed as the remainder of Greek and then Roman civilization fell to the waves of barbarian invaders. Some would certainly have been lost in the destruction of the vast library in Ptolemy's home city of Alexandria, which according to legend was devastated no less than four times.

The library was built at the instigation of an earlier Ptolemy, Ptolemy I, King of Egypt, towards the end of the fourth century BC. It became a unique centre of learning, with over half a million scrolls of information stored in its huge halls. But in 47 BC, Julius Caesar was holed up in Alexandria during the civil war with Pompey. A fire, started to destroy the Egyptian fleet, accidentally spread to the library and burned it down. That time many of the scrolls were saved, but it was then made an intentional target of destruction, twice attacked by Emperors of the collapsing Roman world and finally obliterated by the Caliph Umar, the second of the Muslim rulers to succeed Muhammad.

Around AD 630, the Caliph is said to have ordered that all books in the library that weren't in agreement with the Qur'an should be destroyed. For good measure, any books that *did* agree with the Qur'an were also to be destroyed, as they merely provided unnecessary repetition. The scrolls were piled into the furnaces that powered the battered remains of the Roman heating systems and baths. Ibn al-Kifti, a later Arabic writer, notes in his *Chronicle of Wise Men* that 'the number of baths was well known, but I have forgotten it', (contemporary records suggest around four thousand). According to al-Kifti it took these thousands of furnaces, 'six months to burn all that mass of material'.

Luckily there were other libraries, and some books survived. Later Caliphs were less intolerant of Greek learning; many of the Greek books that we now have, including the remaining fragments of Ptolemy's *Optics*, come to us in translation from Arabic copies made after the fall of Rome. A new force had taken over the burden of knowledge from the ancient civilizations. It took an Arab philosopher who had studied the Greek texts to bring light out from the shadow of the Dark Ages. But first he had to avoid painful death at the hands of the Egyptian king.

Chapter Three

Out of the darkness

There are two ways of spreading light; to be
the candle or the mirror that reflects it.
EDITH NEWBOLD JONES WHARTON

The Dark Ages were dark indeed. Those intent on survival had little time for scientific enquiry. When there was
peace, the Greeks' preference for pure thought was joined by
a Christian distrust of the pursuit of knowledge. A few islands
of advancement remained. Islamic culture, emerging in the
seventh century AD, required every Muslim to pursue knowledge (*ilm*) in order to further justice. After an initial 150 years
of bloody conquest, Islam reached a more peaceful period,
where this pursuit of knowledge brought light to the fore. In
the West, a handful of the inward-looking academics were prepared to risk everything in a stubborn desire for knowledge.
The candles of scientific discovery were few, but they burned
brightly.

The safety of madness

The new Islamic intellectual hub was Baghdad. Here scholars
uncovered the remains of Greek natural philosophy and added
to it with their own researches. In the story of light one name
stands out – Abu Ali al-Hasan ibn al-Haytham, usually called

Alhazen in Western texts. He was born in Al Basrah (now Basra in Iraq) in 965 and was one of a handful of men who would transform the subjective classical theories to devise the geometrical view of optics that we take for granted today.

While young and amenable, Alhazen tried to understand the world from the purely religious viewpoint of his teachers, but he could not resist the more practical approach of dealing with the 'how' rather than the 'why' of the world's workings. His practicality was almost his undoing. Soon famous in the Muslim world as a prodigy who could solve any problem, he was invited to Cairo by the king, al-Hakim. This was the sort of invitation it was not wise to decline. The king had a problem – the Nile. Throughout the history of Egypt, the great river had proved both a blessing and a curse. Depending on its flow, the Nile could ensure plentiful crops, flooding or drought. Al-Hakim commanded Alhazen to devise a means of controlling the Nile, of taking the power of devastation away from nature.

The young Alhazen jumped at the challenge, but soon found that he had bitten off more than he could chew. The Nile did not respond to his bidding. Alhazen was devastated. The king was not interested in failure. If Alhazen admitted his inability to control the great river he would lose more than his job. Fearing for his life, Alhazen cast around for a way out. He considered running away to Syria, but though al-Hakim's kingdom was puny in comparison with dynastic imperial Egypt, he was still a power to be reckoned with. His arm was long, particularly when he was angry. With time running out for his potentially fatal audience with the king, Alhazen made a desperate decision. Tearing at his clothes, a wild light in his eyes, he imitated the madmen he had seen raving in the city streets.

In brief moments of clarity he let it be known that the challenge of taming the Nile had driven him insane. Al-Hakim could not take his revenge on a madman. Alhazen was forced to keep up this pretence for years until the king died.

It's easy to imagine Alhazen, confined in his pretended madness, staring out from a small, barred window at the way light moved and changed in response to the procession of the clouds and the Sun. With little more to do than watch, it became more and more obvious that light came not from the eye, but from the Sun. He noticed that after-images of brightly-lit objects remained floating in front of him when he turned away and looked back into the darkness of his cell. This surely was something external acting on his eye, rather than a response to light that originated in his own eyeball. Similarly the pain he felt when glancing at the Sun could not be a result of light that came from the eye itself. Alhazen convinced himself that light was independent of the eye.

Again, it is easy to imagine that the sight of a busy square outside his window, flooded with sunlight, was the source of Alhazen's other great contribution to the understanding of optics. The cheerful bustle of activity must have seemed like paradise from his silent confinement. Perhaps there were children, scurrying around, yelling and throwing balls to each other. Now that he understood that light didn't require the eye, Alhazen could bring in Euclid's elegant straight line geometry, imagining the light flowing along lines from the Sun, beaming out in all directions. Some of those beams were hitting the square, filling it with brightness – but just one, the one that passed in a straight line into the eye enabled sight. He would have seen the glitter of reflection off mirrors and weapons and metalware in the square. When this happened, he imagined a beam of light hitting the mirror and glancing off the surface at the same angle at which it had arrived, just as the balls of the children bounced off the ground and the walls.

Out of confinement

Once free of feigned madness, Alhazen could refine his ideas. Instead of catching glimpses of reflections in the square, he could work with the polished metal of mirrors. These most simple of optical devices are known to have existed as far back as the Copper Age. They fascinated Alhazen as much as they had his Greek counterparts, and he took the study of reflection to a new peak, detailing the way that rays bounce off different curvatures of mirror from spherical to conical by painstakingly following the paths of hundreds of individual rays. He even used his understanding of light to estimate how thick the atmosphere was.

In Baghdad the sunsets can be spectacular. After the Sun has disappeared below the horizon, the temperature begins to drop quickly, encouraging observers off the roofs of buildings and into the warmth inside. Whenever there was a clear sunset, Alhazen would stay out, timing the twilight, the period of time after the Sun passes out of sight but when faint sunlight still stains the horizon. In his book *Mizan al-Hikmah* he rightly assumes that the light continues to be seen by the same bending that takes place when light passes from air to water – refraction. By estimating how far the Sun would have dropped below the horizon before darkness fell, and combining this with a guess of the amount the air would bend the light he came up with a thickness of the atmosphere of between 15 and 40 kilometres, an impressively accurate result.

Alhazen returned to refraction when he was working painstakingly through the surviving writings of the Greek philosophers that he had managed to scavenge. He was impressed with Ptolemy's experimental approach to refraction, but found when he repeated his Greek predecessor's experiments that it was impossible to get the same results. Though Alhazen had not got the mathematical tools to work out the real relationship, he was sure that Ptolemy's simple constant proportion of the angle the light arrived at and the angle it continued at in water or glass was wrong. However, he still made the impressive guess, not proved until the nineteenth century, that the reason this bending takes place is because the light flows less easily – is slower – in the denser material.

Alhazen had more success with the camera obscura, turning Mo-Tzu's novelty into a working tool. The principle of the device is simple. A pinhole is made in a sheet of dark material that screens off the light coming into a room. The light from the hole falls on a wall opposite, and projects onto its surface a living, moving picture of what is going on outside – only upside down. Alhazen understood this inversion, thanks to Euclid's straight lines of light. The light from the top of an object could be followed in a straight line through the pinhole to end up at the bottom of the image on the wall. Similarly the light from the bottom traced a path to the top of the image.

Later developments of the camera obscura used lenses to turn the picture the right way up, or twisted it through 90° to display it on a table in the middle of the room, but the device remained popular up to Victorian times, used both as an entertainment and to help less than adequate sketchers to produce an excellent reproduction of a scene by tracing the image on a piece of paper.

Not only does the camera obscura provide a useful laboratory for experimenting with light, it illustrates how both the eye and the modern camera work ('camera' is a shortening of camera obscura) – though such linkages would have to wait hundreds of years to be made. For Alhazen the camera obscura was a clear proof of the incorrectness of an obscure Greek theory that husks were peeled off the objects viewed and flowed to the eye. How, he argued, could the husks from a row of candles pass through this tiny hole and rearrange themselves on the other side? Euclid's straight lines and Alhazen's multiple rays from a point were displayed triumphantly in the camera obscura's faint image.

To Western shores

In the thirteenth century, Alhazen's *Kitab al-Manazir*, the summary of his work on light, was translated into Latin, reinforcing a new interest in the phenomenon that had grown from a fresh awareness of Greek philosophy. The newly available Greek texts, translated from Arabic copies, had produced a schizophrenic attitude in the West. Greek natural philosophy was held in great reverence, while any consideration of their non-scientific writing was dismissed as pagan nonsense. It fell on two English clerics to assist the practical rebirth of scientific interest in the West. Both spent time at Oxford University, and though the two may never have met – the older man, Robert Grosseteste, left to become Bishop of Lincoln before the arrival at the university of the second, the remarkable Roger Bacon – it is obvious that theirs was a single strand of thought.

Grosseteste, born in the last quarter of the twelfth century, was a Renaissance man before the Renaissance was ever conceived. He brought a breath of fresh air to a stuffy and rigid hierarchy. He was a campaigner for church reform, sacking abbots for not providing enough priests to look after the people, and constantly criticising the religious abuses of the time, giving a lead that would eventually lead to the formation of the Protestant churches. But Grosseteste was not just a churchman. He was fascinated by music, known for his sharp wit and his deep interest in the natural sciences, and particularly absorbed by the nature of light. It seemed to Grosseteste that light lay at the centre of creation.

In a landmark book, *De Luce*, Grosseteste describes how he envisaged matter to have been formed from light. His ideas combined Plato's Greek philosophy and the additions of Arabic thinkers like Alhazen. Grosseteste's vision was not a detached scientific one as we would recognise it today. For him, as for the builders of Stonehenge or the ancient Egyptians, there was no separation of science and religion. Light's physical characteristics were nothing more than a poor reflection of the spiritual reality. However, Grosseteste's was not a philosophy based purely on faith. He embraced the significance of mathematics and the need to have an experimental basis when acquiring knowledge, setting the scene for the development of science in the future.

The pragmatic attitude, ranking observation above theory, that Grosseteste championed, was contrary to the authoritarian ideas of both the Greek philosophers and the church. But Grosseteste always seasoned his rebellion with political awareness, staying within the bounds of accepted behaviour. Roger Bacon, his philosophical standard bearer, had no such self-control. In an explosion of originality, Bacon brought the scientific method into full flower.

Doctor Mirabilis

Friar Bacon became a legend after his lifetime, making it difficult to separate the truth from the fantasy. Known as Doctor Mirabilis (the miraculous doctor) for his breadth of knowledge, his exploits were embroidered with great enthusiasm. He was said to have created a marvellous brass head that could speak like a person. During his long imprisonment in solitary confinement he was supposed to have educated poor peasants by shouting his teaching through a hole in the prison wall. He was said to be a saint... or to have signed a pact with the devil. What is certainly true is that Bacon's magnificent scientific insight – arguably rivalling Newton or Einstein – was matched only by his consistent bad luck.

For a man of his time he was well travelled. From his birth in Ilchester, Somerset, in 1214 or 1220 (the only source for the date is his throwaway remark when middle aged that 'forty years have passed since I first learned the alphabet', which could have referred to his starting school or university) he had moved to the alien environment of Paris, first learning and then teaching at the university. While there he showed no great interest in science, but when he moved back to England, to the University of Oxford, the nearest Britain had to Paris's great seat of learning, the legacy of Robert Grosseteste's originality opened Bacon's mind to natural philosophy.

Bacon stayed at Oxford for ten years, teaching and studying a heady mix of physical reality and alchemical supposition. At the end of that time he was forced out of the university. This has sometimes been attributed to poor health, but he was already a member of Friars Minor – a Franciscan monk – and it is more likely that his philosophy was too near the knuckle for the senior members of the order. His fascination with the mechanisms of creation was considered very unhealthy. It bordered on magic. And Bacon's pigheaded insistence on applying the same filter to religious teaching as he did to scientific matters, accepting what he found logical but questioning the rest, was dangerously close to heresy. To make things worse, the Franciscan order was already in turmoil.

A visionary Spanish writer, Joachim of Flora, had predicted that a new age of the Spirit would begin in 1260, and that this fundamental transformation of the world would be ushered in by an order of monks, headed up by Merlin himself. Tempted by the scent of power or driven by true faith, some of the Franciscans took the prediction to refer to their order. There was turmoil between the monastic hierarchy and Rome before the church authorities took action, but when action came it was decisive. The Minister General of the Franciscans was ousted for supporting Joachim and replaced by a hard-minded theologian, Bonaventura.

Though the new head of the order was said to be a friend of the scientifically minded priest Albertus Magnus, one of Bacon's teachers and scholastic adversaries in Paris, Bonaventura felt that an unhealthy interest in the outside world had allowed the Franciscans to reach their present perilous state. Among a welter of draconian new rules, Bonaventura forbade monks from writing books or even *keeping* books that weren't first authorized by the Minister General. Roger Bacon was an obvious target for the new regime. Suspected of Joachimite tendencies and teaching magic, he was removed from the university and called back to the friary in Paris to undertake an endless round of menial duties.

The unquenchable candle

If the authorities intended to suppress Bacon, they reckoned
without his immense personal energy. The daily burden of
cleaning and drudgery was not enough to keep him quiet.
Despite the proscription, despite the apparent sin, Bacon was
determined to write. He scraped together some paper and
began to scribble away. Before long he was summoned to give
an account of himself. In all innocence, Bacon explained that
he was working on a series of tables that allowed the proper
calculation of the date of Easter. No one could take this as
heretical, and it wasn't technically a book, so he was allowed
to continue. Under the cover of table after table of numbers he
managed to jot down some first thoughts on a great scientific
treatise he imagined writing – the *Communia naturalium* – and
smuggled these to friends outside the friary to keep them safe.

Then came a minor breakthrough. Bacon managed to contact Cardinal Guy de Foulques, a nobleman who had shown
an interest in Bacon's work before the Franciscan was put on
corrective discipline. This was the nearest he had to a friend in
a high place, and Bacon made the most of it. He pointed out
to de Foulques how much he had appreciated the cardinal's
interest in his scientific work, and explained that he was eager
to assemble his ideas in written form if only de Foulques could
release him from the Minster General's ruling against writing
books.

Although de Foulques was an important man in the church,
Bacon knew his appeal to the cardinal was a long shot. The
Franciscan order maintained a totally separate rule from the
church as a whole, and it was down to Bonaventura, already
shown to be anything but friendly to Bacon's ideas, to make any

exceptions. But de Foulques would certainly have influence, and it was quite possible that Bonaventura would want to help cement his new position by indulging the cardinal. In the long wait until he got a reply, Bacon might have imagined many outcomes, but never the one that finally occurred.

Two years after Bacon wrote his petition, a letter from de Foulques came through. It seemed robust enough to stand up to any opposition from the Franciscan hierarchy. Guy de Foulques, Cardinal Bishop of Sabina and Papal Legate to England, demanded that he be sent immediately, despite any restrictions of the order, Bacon's masterpiece on science. There was one immense snag – there was no masterpiece. Somehow, de Foulques had confused Bacon's plea for support to help him write a book with a claim to already have produced one. De Foulques wanted the book, and he wanted it now. Bacon's nightmare didn't end there. The whole point of writing to de Foulques was to overcome Bonaventura's prohibition, to be able to research the book openly and to get the finances he would need to buy books and paper and services of copyists. Instead, de Foulques, aware of the limitations of his power, ordered Bacon to send his book in secret, without seeing fit to provide any cash.

Bacon was horrified. He could not ignore de Foulques' command, but it seemed impossible to carry it out. He called in all the hidden writings that he could from his friends and began to cast about for a source of funding, all the time trying to keep his activity covered by his harmless numerical calculations. As time went by it seemed less and less likely that he could fulfil his obligation. But then came news that was even more stunning than de Foulques' original letter. The Pope had died and a newcomer, Clement IV, had ascended to the throne

of St Peter. The original name of that new Pope was Cardinal
Guy de Foulques.

Opus Majus

Bacon, by now well used to bending the rules of the order to
suit his needs, wrote to the new Pope, explaining his delay in
responding to the original command and outlining the difficulties he was under given the restrictions placed on him by his
superiors. Despite his often hot-headed enthusiasm, Bacon was
a good enough diplomat to realize this would be a very dangerous letter if it fell into the wrong hands. He worded it with
immense care so there seemed no criticism of the Franciscans
and Bonaventura, and made sure that a close confidant was
given the task of carrying it to Rome. Then, once more, came
the waiting.

It seemed to go on forever. It was over a year before a reply
came, but when it did, to Bacon's relief, it was positive. Taking
the papal throne had not dulled de Foulques' interest in science. But once again there was a catch. In fact, the exact same
catch. There was still no order that Bacon be released from his
penitential labours, no special permission from the Pope to go
around Bonaventura's restrictions. And no money.

Once more Bacon faced a painful dilemma. The only way
he could obey the Pope in producing the book (which after all
he so dearly wanted to do) was to disobey him on the matter
of secrecy. Realizing that he would put himself at great risk,
it took days of deliberation to act. Eventually Bacon took the
Pope's letter to the head of the friary, the Father Superior.
As he had thought, he was immediately placed under suspicion. It was clear from the letter that he had broken one of

Bonaventura's rules by contacting the Pope without clearance through the hierarchy. But Bacon's gamble paid off. After consulting with the local leader of the order, the Father Superior decided that providing the Pope with the book he desired was more important than any minor transgression. Bacon was given the go-ahead to start writing.

The news was exhilarating. It had seemed quite possible that Bacon would never again be able to speak to others about natural philosophy, though every moment his ideas fought inside him in an urgent effort to escape. Now, for the first time since being removed from the university, he could work openly, freely. He was given leave to raise some money, and between efforts at financing the project the words flowed out onto the page like a dammed river that had burst through its restraining banks, impossible to control.

Bacon first intended to write a short letter to the Pope, little more than a brief outline of his new book, but he found it impossible to hold back the ideas. The urge to write was unstoppable after so much mindless manual work. His so-called letter reached a total of half a million words, more than five times the length of *Light Years*. Resignedly, Bacon sent it off to be copied and started on a covering letter for what had now become a major manuscript, but again he was carried away. By the time he was finished he had completed three volumes. The original, the *Opus Majus*, the second, the *Opus Minus*, and the *Opus Tertium*. Between them they covered philosophy, astrology, astronomy, geography, optics, mathematics and more. It was a remarkable, one-man study, and yet it was, as far as Bacon was concerned, only the beginning.

In the final version of his covering letter he encouraged the Pope to commission a new work to cover the whole of scientific knowledge of the time, each part written by the acknowledged expert in his field – an encyclopaedia of science, but aimed at the layman, to bring anyone with access to books into the new light of knowledge. It seemed that there was no limit to what Bacon could do. An unparalleled scientific revolution might have burst forth in the fourteenth century. But Bacon's unfortunate luck was about to take another turn for the worse.

As he waited for a response from the Pope to his proposals and his three works, he was summoned to see the Father Superior. The head of the friary was a little vague, but his message seemed positive enough. Bacon was told that he was to return to England, to Oxford. It seemed, then, that the Pope was satisfied with his work. For a moment Bacon felt pleasure. But he needed to know more. When he asked if the Holy Father had made any specific comments he received news that would pitch him from joy into despair. The Pope had never seen the *Opus Majus*, the *Opus Minus* and the *Opus Tertius*. Before they had reached Rome, Clement IV had died.

Certain suspected novelties

For a while, in the chaos that followed Clement's death, it seemed as if Bacon had no real problems despite losing his protector. He returned to teaching at Oxford and worked on a mathematical treatise and the first elements of his great overview of science. But a new head of the order, Jerome di Ascoli, had been elected. Jerome hated the Joachimite schism (technically not a heresy, or it would have been a matter for the Inquisition) with fanatical zeal. According to a chronicle written around 1370, 'at the advice of many friars the Minister General, friar Jerome, condemned the teaching of friar Roger

Bacon... as containing certain suspected novelties, on account of which this Roger was imprisoned'. While this was written too long after the event to be entirely trusted, he may well have been locked away, perhaps at the priory at Acona where other schismatics were held, kept in chains without human contact until they died, and then to be buried without Christian rites.

The duration of the imprisonment, like everything else in this period of Bacon's life is uncertain. Estimates range from two to 13 years. But however long the isolation, resolute faith and scientific curiosity seems to have kept Bacon going for his long spell in terrible imprisonment. The figure of 13 years is a likely one, as by that time Jerome di Ascoli had become Pope and the new Minister General of the Franciscans, Raymond de Gaufredi, a more sympathetic figure, is likely to have released Jerome's prisoners. It was during his time in prison that Bacon was supposed to have taught simple villagers through a crack in the wall, a fragile contact that might have kept his agile mind sane.

Bacon was returned to Oxford. His books had all been suppressed by the church and remained so, but he was allowed to write again, now confining himself primarily to theology, though enough of his old self came through to show that his views had not changed in exile. He died in 1294 at the age of 80.

Bacon's career was remarkable. It would have been so even if he had only collected the knowledge of the period together, but he was as much an experimental scientist as an encyclopaedist. In fact it was Bacon who made real Grosseteste's scientific principle of putting forward a hypothesis and then testing it against experiment, which holds true to this day. It was a world apart from the Greek philosophers, whose inward-looking methods were not only mandatory at the time, but remained so for another 400 years.

Light was of special interest to Bacon. Taking his systematic approach to science, he picked up on Robert Grosseteste's work on lenses. He showed that they could be used to improve vision (and amongst the Bacon legends, he was said to have made the first ever pair of spectacles for himself). He even leaves hints that he may have constructed an early telescope and microscope, long before the acknowledged date of their construction. He comments in his *Opus Majus*:

> The wonders of refracted vision are still greater; for it is easily shown by the rules stated above [*demonstrating the workings of lenses*] that very large objects can be made to appear small, and the reverse, and very distant objects will seem very close at hand, and conversely. For we can so shape transparent bodies, and arrange them in such a way with respect to our sight and objects of vision, that the rays will be refracted and bent in any direction that we desire, and under any angle we wish we shall see the object near or at a distance.
>
> Thus from an incredible distance we might read the smallest letters and number grains of dust and sand owing to the magnitude of the angle under which we viewed them…

Piece by piece, Bacon combined mirrors and lenses, using his knowledge of geometry to build intricate mathematical tracings for the paths of light rays. He argued forcibly against Aristotle's widely held view that light could not take any time to travel, suggesting that it was in some ways similar to sound.

He described how the rainbow could be caused by refraction and reflection within individual drops of rain. He brought a fresh mathematical view to Alhazen's work on refraction. And he gave a reasonable explanation of why the Sun and Moon look larger near the horizon.

Bacon's contributions to optics were significant, but most of all he acted as midwife for science itself, helping it into birth as a separate discipline from philosophy. Bacon was arguably the first true scientist. Not a good scientist, admittedly. He still placed too much emphasis on the word of authority, and often confused experiment with experience and word of mouth. But it would be very surprising if the first in the field had been particularly good at it. Given the incredible odds he worked against, he deserves a brighter place in the scientific galaxy than he holds today.

A new perspective

On a bright day around 1420, a remarkable man stood outside the great cathedral of Santa Maria del Fiore in Florence. This was Filippo Brunelleschi, who was to have a huge impact on the development of painting, an impact dependent on his understanding of the behaviour of light. Brunelleschi had a great advantage over his artistic contemporaries – he was not a pure artist. As an architect, he was an enthusiast for classical design who had a feel for the mathematical rightness of form that was alien to many of his brethren. Used to working in three dimensions, he knew there was something visually wrong in the flat, unnatural paintings of the time.

Brunelleschi had not always designed buildings. As a metalsmith he entered the great competition of 1401 to design the bronze doors of the Florentine baptistery. Failing there, he moved on to the bigger canvas of buildings themselves. But he retained an expertise with devices from his metalworking days. Unable to persuade his contemporaries that there was anything wrong with their paintings, Brunelleschi decided to construct an engine that would prove the superiority of his vision. If the world would not listen, it would have to be shown.

Brunelleschi is often described as a young man on the day that he transformed artistic vision, but the best estimates for his demonstration place it between 1415 and 1425, which would have made him anywhere between 38 and 48, already well experienced in the accuracy of eye needed to be a great architect. In his hands as he stood on the cobbled verge of the Florentine piazza was a simple device, yet one of wonderful ingenuity. It was a sales tool that would have made any twenty-first century marketing manager proud. With a mirror, a piece of board and a small hole, Brunelleschi was ready to demonstrate the value of perspective.

He had painted a mirror-image view of the baptistery at the far end of the piazza. Brunelleschi combined his architect's awareness of spatial form with Euclid's understanding of the way light flowed in straight lines from the object seen to the eye. If he followed these straight lines back from the eye to a series of identical objects that were further and further away, each subsequent object would seem smaller, along the straight line beam of light. Similarly, anything in his picture that receded into the distance had to get smaller in the image. He had brought perspective into the view.

On its own this was impressive, but Brunelleschi knew that he would need a more effective sales pitch to overcome an ingrained view of how paintings *should* look. He bored a small hole through the middle of the painting and turned it round, so that the viewers saw only the bare back of the board. Each held it up to his or her eye, peering through the hole at the baptistery. Above the great building, the sky was a rich blue, peppered with white clouds in majestic motion. Now, as the observer continued to look through the hole, Brunelleschi blocked the view with a mirror. Instead of the building itself, the mirror showed the reflected painting on the back of the board. To the surprise of his friends, they saw an identical view. The painting was a true reflection of reality.

The final touch, the masterstroke of illusion, was that Brunelleschi had not painted the sky in his picture blue, but instead had made it of polished silver, so it reflected the real sky with its drifting masses of cloud. The impact was remarkable. Until then there had been no more reality in paintings than there was in a child's grotesque drawings. Artworks reflected the emotional impact and significance of the components of the picture, rather than the optical reality of vision. Now a painting could be a mirror for creation. It's interesting that Brunelleschi's architectural style during the same period was dominated by geometrical form, his fascination with sight lines and geometry stretching even into the great buildings of Florence that he designed.

It would be unfair to describe linear perspective as being entirely Brunelleschi's discovery. As far back as the Greeks, some elements of perspective had been understood, and others, notably his contemporary the painter Masaccio, had successfully incorporated perspective elements into their work – but Brunelleschi's combination of the vision to see the value of a true perspective view with the showmanship to come up with this outstanding demonstration set him apart. By the time

Albrecht Dürer had produced his *Painter's Manual* a hundred years later, an understanding of linear perspective had become such an essential component of the painter's skill base that Dürer concentrated on geometrical form in depth before considering any aspect of actually putting paint on canvas. Dürer suggested the use of a grid to aid the eye in correctly assessing perspective. Brunelleschi's architectural viewpoint had come to dominate art.

Art and the eye

There was certainly no doubt that perspective was a familiar tool to another great artist-scientist – perhaps the man who best combined the two in all of history – Leonardo da Vinci. Other artists would turn to science, and many scientists still dabble in the arts, but Leonardo excelled in both fields. His first love, though, was art. Science was to come into his portfolio for eminently practical reasons.

Leonardo's talent flowered early. There was no reason why a fourteen-year-old from a rich family in Florence (the family moved from Vinci while he was still young) should take on an apprenticeship unless he already had a special talent, but this Leonardo did in 1466, under Andreo del Verrochio. For 12 years he was to work in the studio of this master painter and sculptor before launching his career with a first commission to paint an altarpiece for the Florence town hall. Like many of his works this was never finished.

Leonardo didn't stay in Florence long once he was independent. Instead he wrote what must be one of the most deceptively self-flattering letters ever to the Duke of Milan, claiming to be a military and nautical engineer of unrivalled excellence (oh, and by the way, he sculpted and painted a bit too). Leonardo, who got the job on the basis of this wonderful piece of bluffing, must have already had an interest in things scientific, but as the Duke's principal engineer he showed amazing flexibility in the speed with which he picked up technical knowledge.

He couldn't settle. Within eight years he was shuttling between Florence and Milan with all the nonchalance of a modern transatlantic businessman. After a period with the Pope as his client, he finally travelled to France to work for King Francis I, and died there in 1519. Throughout his career, whether employed as an artist, engineer or scientist, Leonardo continued to paint and to explore scientific concepts, but often his work was incomplete. It's not clear whether this was a result of a butterfly mind that always kept him skipping on to the next challenge, or the outcome of frustrated perfectionism. Perhaps it was a little of both.

Sometimes in Leonardo's diaries and notebooks there are intriguing hints that he may have made much more progress with light and optics than has ever been confirmed. Two brief fragments suggest that he was experimenting with the technology of telescopes a hundred years before they were thought to have been invented (though nearly 200 years after Roger Bacon's similarly tantalizing but more detailed comments). Leonardo writes:

Make lenses to see the moon big.

and

In order to observe the nature of the planets, open the roof and bring the image of a single planet onto

> the base of a concave mirror. The image of the planet
> reflected by the base will show the surface of the planet
> much magnified.

providing an intriguing glimpse of a discovery that was never
fully recorded.

Because of his interest in physiology, and a conviction that
vision provided the ultimate record of experience, Leonardo's
biggest clear contribution to the history of light was in understanding the human eye as none really had before him. He
came to the conclusion that the eye worked in a similar way
to a camera obscura, projecting an upside-down image on the
back of the eyeball, though he made no attempt to experimentally prove this idea, or to explain why we don't see the
world upside down. Leonardo never grasped the full picture
of the eye's optics. It was a fellow countryman, the Benedictine
monk Francesco Maurolico who a few years after Leonardo's
death suggested that the double lens structure of cornea and
lens focuses the rays of light on the retina, the nerve at the
back of the eye, from which it is conducted to the brain.
Maurolico's theory enabled him to explain that both near- and
far-sightedness resulted from having the incorrect depth of eye
to correspond to the focusing point of the lenses. Maurolico
developed the theory behind Bacon's eyeglasses, but it was
Leonardo that pointed the way.

The new Universe

One figure remains in this period when no great breakthroughs
were made in our understanding of light, but a few isolated
individuals laid the foundations on which the next generation

would explode so spectacularly into view. This last contributor was Nicolaus Copernicus.

It was his insight that rationalized the structure of the Universe, and would later condemn Galileo to imprisonment. When Copernicus (more properly Niclas Kopernik, but he has become universally known by his Latinized name) was born in the Polish town of Thorn (now Torun) in 1473, there was no doubt about the structure of the Universe. It was just as Aristotle and Ptolemy had imagined it so long before. At the centre of everything was the Earth. Round this rotated the Moon, the Sun and the planets, all contained by the great circuit of the stars. Each of these levels was held in place by immense crystal spheres. Forgetting for a moment these unlikely spheres, it made a sort of sense. It didn't take a genius to see that the celestial bodies, including the Sun, revolved around the Earth; you only had to watch the skies. Concepts like sunrise and sunset were deeply embedded in human culture. But Copernicus was not prepared to fall in meekly with the accepted wisdom of the day.

Copernicus seemed incapable of sticking with a single subject while at university. He showed distinct enthusiasm for remaining an eternal student. After four years at the University of Krakow studying liberal arts he didn't get a degree but instead moved on to Bologna to study canon law. In Bologna he lived at the house of a maths professor, Domenico Maria de Novara, whose enthusiasm for hands-on science soon began to rub off on the young Copernicus. Together they studied the heavens, and Copernicus was bitten by the astronomy bug. Within a couple of years he was so fired up that he was prepared to give a lecture on the subject in Rome when visiting the Holy City with his brother to celebrate the jubilee of 1500

– though it's hard to believe that he was much of an expert at this stage.

Astronomy remained important to Copernicus, but he had no intention of letting it get in the way of his wider education. He had already been elected a canon of the chapter of Frombork (Frauenburg), a shrewd move by his uncle as it made sure that Copernicus had a basic income with little responsibility. In fact he would receive a stipend for over 20 years before he actually took up the position. Copernicus moved on from Bologna to Padua to study medicine, but as with his liberal arts studies at Krakow, he never received a degree. When he finally did take a doctorate it was in canon law, the subject most relevant to his chosen career. Strangely, the degree was awarded by a university he had not attended – Ferrara – but it was common practice at the time for students at a prestigious university to go elsewhere to receive their degree. They could still claim to have studied at the right place, and the qualification was a lot cheaper.

Back home in Poland, Copernicus lived with his uncle, Bishop Lukasz Watzenrode in his palace at Lidzbark Warminski, putting his canon law training to good use in the administration of the diocese and moonlighting as a part-time doctor. Despite his incomplete medical training he was popular both as a court physician and with the poor, who he treated without charge. It was either there or shortly after his uncle's death when he moved to Frauenberg that he wrote his first book criticizing Ptolemy's Earth-centred Universe. The lengthily titled *De Hypothesibus Motuum Coelestium a se Constitutis Commentariolus* did not stir up any great reaction. It was at least another 15 years before he finally penned his masterpiece *De Revolutionibus Orbium Coelestium*, and even then the book was not published for a further 13 years, printed days before his death on 24 May 1543. It has been claimed that he held back the book until he was dying, because he knew the subject would be dangerously controversial, but Copernicus is unlikely to have been this devious. Even so, the controversy was real enough.

Copernicus moved the Sun to the centre of the Universe, leaving the Earth (spinning on its axis every day) revolving around it with the other planets. He didn't throw out everything that Ptolemy had set out. The planets and stars, for example, were still supported by their immense crystal spheres. But the Copernican change was fundamental. So much so, that a common reaction at the time was to miss out the essential part – that the Earth rotates around the Sun – and pick out for use any of the remnants that seemed practical and sensible.

The fascinating mystery is exactly how Copernicus came to this revolutionary theory. Admittedly it does overcome many problems that Ptolemy's system caused. If everything really rotated around the Earth, there was no obvious reason for the cyclical progress of the seasons through the year. (This is actually caused by the tilt of the Earth's axis, but would be later used as argument for the Earth's motion.) And forcing the planets to revolve about the Earth required some very nimble footwork to explain why they occasionally change direction in the sky. It also seemed strange that Mercury and Venus, the two inner planets, never strayed far from the Sun. All this made so much more sense with Copernican theory. But it remained a huge leap of imagination.

Part of Copernicus's inspiration was observation. He spent much time shivering in the cold night air, perched in the tallest towers of the town, studying the motion of the planets with the crude wooden instruments that were the best available at the time. The very first spark of his idea, though, could have come from antiquity. Although the Greeks had generally put the Earth at the centre of the Universe, there was one dissenter, Aristarchus. His writing on the subject was lost with so much else in destruction of the Alexandrian library, but we do know what Archimedes thought of his theory. In a rather strange little book called *The Sand Reckoner*, in which Archimedes sets out to calculate the number of grains of sand it would take to fill the Universe, he wrote:

> [...] the universe is the name given by most astronomers to the sphere the centre of which is the centre of the Earth, while its radius is equal to the straight line between the centre of the Sun and the centre of the Earth. This is the common account as you have heard from astronomers. But Aristarchus has brought out a book consisting of certain hypotheses, in which it appears, as a consequence of the assumptions made, that the universe is many times greater than the 'universe' just mentioned. His hypotheses are that the fixed stars and the Sun remain unmoved, that the Earth revolves about the Sun on the circumference of a circle, the Sun lying in the middle of the orbit, and that the sphere of fixed stars, situated about the same centre as the Sun, is so great that the circle in which he supposes the Earth to revolve bears such a proportion to the distance of the fixed stars as the centre of the sphere bears to its surface.

Because all that remains of Aristarchus's remarkable book

is a comment from a hostile critic there is a frustrating lack of context. Where did these original and surprisingly accurate ideas come from? Copernicus would have been familiar with the writings of Archimedes, but it seems unlikely that this tiny fragment could do more than provide the seed of an idea, while the fruition of his theory depended much more on his observation and original thinking.

It might seem equally odd, though, however great the original leap, that it took so long for this idea to be widely accepted. Once the Copernican theory was explored it explained so much that was messy and uncomfortable in Ptolemy's system. The resistance was twofold. In part there was a religious backlash. The Earth had to be at the centre of the Universe, because God had taken such an interest in it. To argue against the Earth being at the centre of everything verged on an insult to the deity. What's more, a pair of references in the Bible seem to support the Earth-centred theory, referring to the Sun 'staying its course' (stopping its motion, so it must be moving, not fixed at the centre of things) and the Earth being 'ever immovable'. Nearly a hundred years after the death of Copernicus, this was a strong enough argument to put Galileo in prison.

Even though *De Revolutionibus* was written with the encouragement of the Catholic Church and dedicated to the Pope, it made some of Copernicus's colleagues nervous. When the book was published, his editor managed to dilute the force of his argument, turning his theory into a 'hypothesis' and replacing the original introduction with one that warns the reader not to expect absolute truths from astronomy, nor to necessarily accept the Copernican view as true. Before long, though, there was increasing discomfort at the content, with the authorities first insisting on injecting qualifications into any sentences that made it seem that the Sun-centred picture was right, and finally clamping down on those who agreed with Copernicus.

The other problem with the Copernican theory was that it ran counter to the views of the ancients, and Greek philosophy was still held in enormous respect. It took the radical thinking of Newton and a handful of others to spread the Copernican view as the inevitable one, initially in England, France and the Netherlands, and eventually through the whole of Europe. This was a slow process, though. It wasn't until the end of the eighteenth century that the Copernican solar system triumphed universally.

By putting the Sun at the centre of things and shifting attention away from the Earth, Copernicus had done more than change the astronomical map; he had made a direct impact on the study of light. It was the understanding of how the Earth and the other planets rotated around the Sun that would make it possible to measure the speed of light 150 years later. More important still, by moving the Earth away from the centre of the Universe he had also, symbolically, moved humanity away from the centre of all consideration. Light, which had always had such subjective associations, could now be studied objectively for what it was. The new age of science that saw its first, tentative steps in Roger Bacon had truly begun.

Chapter Four

Engines of light

I now wish to discuss some principles which belong to
optics. If the consideration just mentioned [mathematics]
is noble and pleasing, the one in hand is far nobler and
more pleasing, since we take especial delight in vision,
and light and colour have an especial beauty beyond
the other things that are brought to our senses.

ROGER BACON

The human eye is an amazing instrument, but it can only see so far. As optical science blossomed, devices to improve on the eye's capabilities were added to the scientists' toolkit. The optical powerhouses of Renaissance science were the microscope and the telescope. In principle these engines of advancement could definitely have been developed earlier. Basic lenses had been around since before Bacon's time, while magnifying mirrors date right back to Archimedes. We have seen that both Bacon and Leonardo may already have dabbled with early forms of telescope and microscope.

Even so, it shouldn't be too much of a surprise that an explosion of instruments to explore inner and outer space took place in the protestant European countries around the second half of the sixteenth century. Copernicus had opened eyes and minds by shifting the Earth from the centre of the Universe. With a transformed understanding of our place in creation it was natural to have a greater interest in just what was out there. At the same time, the technology of lens production had improved sufficiently to make multi-lens instruments practical. The stage was set.

The compound view

The simple action of putting two lenses together in a tube transformed our ability to delve into the realities of microscopic life. Such compound instruments use multiple lenses to increase the magnifying power. In a two-lens microscope, a lens close to the object being studied produces a magnified image on the opposite side of the lens. The second lens, the eyepiece, then acts as a magnifying glass, focused on this already enlarged image.

Compound microscopes were first developed by a father and son team. Hans and Zacharias Janssen were Dutch lens grinders. It is Zacharias that now tends to be remembered, as he built his career around optical instruments, but when the first device was assembled, around 1590, he was only a boy, so Hans probably deserves the bulk of the credit. Another name that is frequently connected with early microscopes is Anton van Leeuwenhoek. It's true that he was responsible for one of the first breakthroughs using a microscope, discovering bacteria in 1674, but there was nothing remarkable about van Leeuwenhoek's instrument, which had only a single lens and so was little more than a powerful magnifying glass on a stand.

Seeing further

The telescope's origins are much less clear than those of the microscope. Even unaided, the eye is a superb long-distance instrument. On a clear, unpolluted night, a candle flame is visible to the naked eye from around 14 kilometres away. Around half a dozen photons are enough to trigger the optical nerve. To see just how impressive that half dozen photons is, bear in mind that a 100 watt light bulb throws out around 100,000,000,000 photons every billionth of a second. We sometimes don't appreciate just how effective the eye is, because of the way it compensates for relative brightness. The light of the full Moon, for example, is 300,000 times weaker than sunlight, but we can still see surprisingly well in moonlight. Even so, when it comes to the far reaches of space, the eye needs help. Just as combining two lenses to make a microscope opened up the miniature world, lenses and mirrors provided the means to bring otherwise invisible distant objects into view.

Telescopes come in two main types: refracting, where distant images are magnified by a series of lenses, and reflecting, where the light rays are collected and the images magnified using curved mirrors. The best known early telescopes were refractors, like that of Hans Lippershey, the man whose name most often appears in the history books as the telescope's inventor. Unfortunately, in this respect at least, most history books are wrong.

There was an amazing boom in the optical business in Holland around the turn of the seventeenth century. For a brief period, less than a hundred years, the Netherlands became *the* centre of optical mastery. The country had never had the same scientific significance before and never would again. This was a time when a particular speciality could be concentrated in a single country or even town. The unparalleled success of Holland's lens grinding businesses meant that there was a unique opportunity to investigate the workings of optics.

Like many other Dutch spectacle makers, Lippershey experimented with the possibilities of combining various lenses, but any credit he has received for inventing the telescope derives from his ability to shout the loudest. Just as the notorious Amerigo Vespucci had America named after him thanks only to his own highly doubtful claims, Lippershey benefited from shameless self-publicity. His attempt to become the man behind the telescope got an early bad press. He put in a patent claim, but two other spectacle makers, Zacharias Janssen and the lesser-known Jacob Adriaanzoon, both claimed to have already made similar inventions. Although they couldn't prove their prior claims, neither could they be disproved and the patent wasn't granted, much to Lippershey's disgust.

In reality, telescopes had probably been built significantly earlier than the wave of Dutch inventions. As we have seen, both Roger Bacon and Leonardo da Vinci had made intriguing hints about the use of lenses and mirrors and may have constructed some early prototypes. But the most likely true inventors of the telescope were the English father and son, Leonard and Thomas Digges. Leonard was an adventurer who had the considerable luck to survive a failed revolution against Queen Mary. Thomas was a noted scholar and the main English supporter of Copernicus at that time.

When Thomas wrote about his father's work after Leonard's death, he claimed that they had used 'perspective glasses' to see distant objects. The military potential of such an invention was not overlooked by Queen Elizabeth's court. Asked to investigate these claims, William Bourne, an expert on military technology, described the telescope and some of its limitations (it had a very narrow field of view), suggesting that there was an actual device for him to examine, rather than just a theory. Painstaking research by Colin Ronan shows that the Digges family had a reasonable claim to have been the makers of the first true telescope. Theirs was a very poor instrument, using a clumsy combination of a lens and a mirror, but it has a significant place in history.

On reflection

One of the problems with refracting telescopes like Lippershey's is that they get longer and longer as they become more powerful. Within 50 years of the popularization of the telescope, leading-edge instruments were 150 feet long, making them painfully unwieldy. The reflecting telescope, which uses a curved mirror to focus incoming light rays onto a point, was destined to overcome this problem. The first known reflector (if we ignore the Digges' hybrid device) was designed by a Scottish scientist, James Gregory. To make a usable reflecting telescope, Gregory had to overcome a practical obstacle. The curved mirror focuses the light rays at a point that lies in the path of the incoming light. If you put your head there to see the result, it blocks the incoming rays and you see nothing. To avoid this, a reflector needs some mechanism to get the light out of the telescope and into your eye. Gregory used a small mirror part way up the tube to send the light back through a hole in the centre of the main mirror and out the back end of the telescope.

But it was Newton, a few years later, who really helped the reflector to take off. His telescope was built to overcome the coloured fringes that refractors produce (because of the way different colours are bent by different amounts when they shine through a prism or lens – see page 86). Instead of sending the light through a hole in the main mirror to view it, he used a small, flat mirror, set at an angle, to reflect the light through ninety degrees and out through the side of the tube. This style of reflector, called Newtonian in his honour, remains popular for smaller telescopes, though large professional instruments more often use a variation on Gregory's design called Cassegrainian after its French inventor. Again the light is reflected back through a hole in the main mirror, but by a close convex mirror, which allows for shorter telescope tubes.

After the frenzy of invention in the seventeenth century, the basic science of the optical telescope has hardly advanced. Though telescopes are much bigger today – for many years the world's largest was at Mount Palomar with a 200 inch (five metres) mirror, but recently instruments have been built up to around 330 inches (over eight metres) – and the telescopes are sometimes sent into space to overcome the problems of working through a distorting atmosphere, the optical technology is often much the same. Modern telescopes can weigh hundreds of tonnes; it is in the computerized drives necessary to move them with minute precision, and the electronic cameras and sensors that have replaced the viewer's eye, that telescope technology has mostly moved on, though an innovative concept called adaptive optics is now being used to overcome some of the problems faced by ground-based telescopes.

Imagine looking down a long, straight road on a hot day. The surface seems to shimmer, turning to liquid – the air above the road dances and ripples. Adaptive optics devices undo the visual warping, stabilizing the image and removing the complex distortions that make it so difficult to get a clear picture of the sky through our atmosphere. Before powerful computers were available this simply was not possible, but adaptive optics uses a range of computer-controlled technology to clarify the image. In some equipment, the light is first bounced off a conventional mirror that can be steered very quickly to change the angle of tilt. This irons out vibrations. The light then hits a second, flexible mirror that can be distorted in shape. A sensor samples part of the light before it hits the mirror, and monitors the position of a set of easily identified points – as these move, so the mirror is distorted to undo the movement. Typically the mirror might be reshaped several hundred times a second.

Outside of visible light much greater advances have been made. Now telescopes are at work in practically every part of light's spectrum. Close relatives to the visible telescopes – those working in infrared, for instance – are very similar to the traditional devices, but others are really only telescopes in name. Radio telescopes are ultra-sensitive aerials, using vast dishes to focus radio waves onto the receiver. For a while the trend was to build bigger and bigger dishes. The Jodrell Bank telescope in England is one of the best known at 200 feet (61 metres) across. It has the flexibility of being steerable to point at any part of the sky, but is dwarfed by the fixed concrete dish at Arecibo in Puerto Rico, which is 1,000 feet (305 metres) across. Such monsters have largely become redundant as arrays of smaller dishes, their results combined by computer, can cover a wider area more cost effectively.

Telescopes have become more than just tools to explore space. The sheer size of the Universe means that light takes billions of years to cross it. By looking into the far distance, telescopes act as time machines that peer back towards the origins of time. Their place in the scientific armoury remains as important now as it was in the 1600s.

Science and finance

Any quiz asking 'Who invented the telescope?' is unlikely to get the answer Lippershey or Digges – the chances are that the invention will be attributed to Galileo. Though Galileo Galilei wasn't the first to design a telescope he combined a particular flair for making use of it with masterful salesmanship. Cash (or lack of it) plays an important part in Galileo's story. The family he was born into in 1564 was not particularly poor, but they had not got money to burn. Galileo's father, Vincenzo, was a court musician in Pisa, home of Italy's famous leaning campanile. Vincenzo's was a respected job, if not highly paid.

Even so, there was money enough for Galileo to be given a reasonable education, with the intention of his becoming a doctor. Galileo went along with this sufficiently to enrol at the medical school of the University of Pisa, but soon made it plain where his interests lay. He became fascinated with mathematics. Vincenzio enjoyed the subject himself, but was determined Galileo should follow a more financially rewarding career. Galileo ignored his father's wishes, continuing to study maths, even though this meant that he ended up leaving Pisa University without a degree.

While at university, Galileo's distaste for the classical methods that were still taught without question grew. The philosophy of the ancient Greeks, and particularly Aristotle, held total sway in academic circles, placing all the weight of argument on human imagination without ever requiring an assertion to be tested. If something was held to be true, in the Aristotelian view, it was true. Because many elements of Aristotle's philosophy fitted very well with that of the church, it could be dangerous to argue with this approach. This was a time when the Inquisition still held real power. Galileo's feelings echoed those of his father, Vincenzo, who wrote that those who rely purely on authority 'act very absurdly', but this was in a book on music that was too low profile to put him, the elder Galilei, in danger.

To Galileo, experimentation was at the heart of science. When he wasn't inventing devices to keep the cash flowing – most notably in feats of military engineering – he was forever tinkering and observing. He took the position of chair of mathematics at Padua to formalize his work, but found his new status a mixed blessing. The job involved a considerable amount of teaching, eating into his research time. Before long, his father's early death added the need to provide dowries for his sisters to the burden of supporting his own growing illegitimate family. Galileo began to search for a post that would give him better financial stability, and the chance to dedicate more time to research. He set his eyes on the position of mathematician to the Tuscan court. This was no random decision. His father had worked for the Duke of Tuscany, and Galileo's first exposure to the joys of mathematics had been through meeting the then court mathematician, Ostilio Ricci.

Galileo got a nose into the court when the Duchess Christina asked him to give some tuition to her son, Cosimo. They got on well, which was to prove a valuable asset when Cosimo became Grand Duke less than four years later. Galileo was doing excellent groundwork, but there was no immediate sign of a post coming up in the court, and his need for cash was pressing. When a classic get-rich-quick opportunity arose, Galileo was not slow to act. His fast, if disreputable, response was to link his name forever with the telescope.

Beating the opposition

As we have seen, the telescope had probably been in existence for 50 years, and had already become the subject of a patent dispute in Holland. Now, news of these amazing instruments reached Southern Italy. Galileo's Paduan base was part of the Venetian Republic, and Galileo, ever one to make sure he remained popular in the right circles, was often in Venice. On one visit he was shown a letter describing the telescope. Immediately the commercial and military potential occurred to him. The sea was essential to the Venetian power base – a working telescope would be a powerful tool for mariners in both peace and war, and worth an impressive financial reward.

Galileo's dreams of financial ease were soon endangered. He heard that a Dutchman, from the home of the early seventeenth century telescope, was in Padua demonstrating an instrument. Galileo rushed from Venice to Padua, but with the precision of a carefully scripted farce, the Dutchman decided it was time to move on to the capital and headed for Venice just as Galileo left. The news that he had missed his man spurred Galileo into hasty action. With his businessman's sense he knew exactly what the Dutchman would attempt to do – sell his telescope to the Venetian ruler, the Doge. All that Galileo had in his favour was contacts in the right places, his technical genius and luck. Each was to play its part.

Galileo began to throw together the various lenses he had available, trying different combinations and mountings. Luck came into play when he tried combining a convex lens (bending outwards) with a concave (bending inwards). The Dutchman was using a pair of convex lenses, which produced an upside-down image. This is fine for astronomy, but irritating when trying to spot ships at sea. Galileo hit on a combination that gave a properly oriented view. Meanwhile, the Dutchman had approached the Venetian court, and here Galileo's contacts came in. A close friend of his, Friar Paolo Sarpi, was asked by the Venetian authorities to investigate the Dutch device. Sarpi managed to sideline the rival telescope maker long enough to allow Galileo to demonstrate his own telescope, constructed in a single day, directly to the Venetian senators.

It was an outrageous success. The elderly senators had to be restrained from fighting over the chance to be the next to climb to the roof and scan the horizon for ships. They were like children with a new toy. At this point, Galileo could probably have named any price for his invention, but instead he showed excellent commercial judgement, presenting it as a gift to the Doge in a handsome leather case. His investment paid off. He was offered a lifetime extension of his position at Padua on double pay – a much better option than a single payment.

At least, better in principle. But some practical difficulties remained. The pay rise would not be delivered until the following year, and staying on at Padua would mean a continuation of the teaching that continuously interrupted his favoured combination of experimentation and social ingratiation. For Galileo, this offer was a valuable fallback, but he had not lost sight of the rich possibilities offered by the Tuscan court.

The Starry Messenger

Not only did he make sure that Duke Cosimo had a chance to play with his next telescope, Galileo was careful to continue to make the Tuscan duke a central figure in his work. Early in the next year, 1610, Galileo discovered the four largest moons of the planet Jupiter, using a new and more powerful telescope. Writing about his discovery in his famous book *The Starry Messenger*, he made sure that his masterpiece was dominated by a lavish dedication to Cosimo. Within a couple of months his astute manipulation of the nobleman's affections paid off. He was given a combined post of chief mathematician at Pisa and court mathematician to Cosimo – on the same salary as he had been promised at Padua, but with cash up front and no teaching duties.

From then on, Galileo's finances were secure and he could concentrate on applying himself to a wide range of topics. While at Padua he had already tried unsuccessfully to measure the speed of light (see page 136). Although he continued to make and use telescopes, it was the motion of the planets and the workings of earthly mechanics that were to fill his attention. He made no further contributions to optics, and his only remaining dalliance with light was to come shortly after getting the Tuscan position. In 1611 he took a mystery box to Rome, showing off the contents in darkened rooms to amazed audiences.

The box contained a mineral, barium sulphide, called *spongia solis* (solar sponge) by its Bolognese discoverer who had brought it back from the Indies. The solar sponge was remarkable because, though cold, it glowed in the dark. Quite what was happening was a mystery, but it led Galileo to think that there was a relationship between the production of light and dividing material into its component atoms, a guess that was remarkably close to reality. On the same visit he was granted an audience with the Pope and made a member of the Lincean Academy, the forerunner of all the other great scientific societies.

Galileo continued his investigation, his writing and his social manipulation for 20 years without incident, but he was always dancing on the edge of disapproval, as everything that he observed confirmed more and more that Copernicus's view of the Solar System was correct. After all, if Jupiter had moons that rotated around it, rather than travelling around the Earth, why should everything else be centred on our planet? By 1630, Galileo was ready to write up this theory, and with characteristic care he made sure that every possible authority from the Pope down was happy with what he was doing. His idea was to write a book in a style that had been common since the Greeks. It would take the form of a conversation between two protagonists, one holding the Copernican view, the other Aristotle and Ptolemy's classical picture.

Galileo passed his book through the office of the official censor and made all the requested changes before publication. But surprisingly for a man with such good social and political skills, he allowed himself a little joke that backfired in a big way. The character presenting the classical view was called Simplicio. Now while there was an actual Greek philosopher of this name, it was hard not to assume that Galileo was labelling those who held the conventional view (that's to say, the church's view) as simple. Worse still, at the censor's request, Galileo had added a postscript emphasizing that the church endorsed the classical model of the Universe. These words, arguably those of the Pope himself, were not left as a neutral postscript but were put into Simplicio's mouth. The usually politically astute Galileo had laid himself open to attack. It was not long in coming. Galileo was charged with heresy.

In a straightforward trial Galileo should have been safe. Not only had his book been passed by the official censor, but he had documentary evidence that the previous Pope had been happy with his discourse as long as it remained hypothetical.

But trials for heresy had a habit of coming out the way the Inquisition desired, and Galileo was found guilty. Before sentencing he admitted that he had gone too far. Instead of his potential fate of being burned at the stake he was committed to life imprisonment, initially at the Tuscan embassy and before long in his home. For his remaining nine years he concentrated on his work, writing up his great theories of motion. He continued to experiment and invent even after going blind, dying in 1642. It was another 350 years before Pope John Paul II gave him a pardon, accepting that Galileo's views were correct.

Practical philosophy

It's curious that Galileo's practical approach to science would be carried forward by a man we now associate almost wholly with philosophy – René Descartes. As a boy, Descartes had little time to experience nature in his hometown of La Haye in the Touraine region of France. From the age of eight, in 1602, this son of an aristocratic family attended the La Flèche Jesuit school in Anjou. From there he went on to study law at the University of Poitiers, but this he intended to be a mere stepping-stone to a military career. Soon after leaving university he entered the service of Prince Maurice, the ruler of the Netherlands.

The realities of army life proved less attractive than the promise. Descartes soon came home to France, where perhaps inspired by his Jesuitical training, he began to take a serious interest in natural philosophy. But while he had not felt any urge to stay with Prince Maurice, the Netherlands proved very attractive. Descartes returned there when he was 32 and stayed for most of the rest of his life. Perhaps the only benefit of his military experience, apart from providing the inspiration for this move was to encourage the very hands-on approach he took to understanding the eye.

Leonardo da Vinci had already suggested that the eye worked much like a camera obscura, and this idea was later backed up by Johannes Kepler, the German astronomer who showed that the orbits of the planets were not circular but elliptical. Descartes proved the reality of the camera obscura theory with a graphic demonstration. He obtained a bull's eye from the abattoir and scraped off the back of the bloody organ. Revealed on the hazy screen he uncovered was an upside-down image of the world. Like Kepler and da Vinci, Descartes could see the resemblance to a camera obscura, but had gone one step closer by proving that an image was projected onto the retina by the lens at the front of the eye.

On a roll, Descartes also made a clear description of the way that light reflected off a mirror, arriving and leaving at the same angle to the mirror, and though this observation can be traced all the way back to Alhazen, Descartes is still acknowledged as the first to write it down explicitly as a law. Unfortunately, his subsequent plunge into explaining light's nature was much less straightforward.

A tendency to motion

Descartes' starting point was thinking through the process of light getting from a distant source like a star to the eye. He envisaged an invisible 'something' that filled empty space, which he called the plenum. Light, he thought, was a 'tendency to motion' in the plenum, resulting in pressure on the eyeball that generated the perceived light. Descartes' theory, while decidedly shaky, is often considered the starting point of the modern science of light, as it involves only the source and the medium through which light is transmitted, something that had never been stated explicitly before.

If Descartes had been right it would have meant that light had to move instantly from its source to the eyeball. It's as if there were a huge invisible snooker cue stretching all the way from a star to the eye that views it. When the star pushes one end, instantly the other end pushes against the eye. In practice, Descartes thought the 'cue', his plenum, consisted of a vast number of tiny, inflexible, invisible spheres. He imagined all of space filled with these minuscule balls. Pressure on one ball would be transmitted through millions of others before reaching its destination. The spheres would act as if there was a single object linking cause and effect. Light, he thought, didn't actually move, but was just something that made motion possible, a bit like the pressure of someone's hand on your shoulder. That push isn't itself motion but something that encourages motion – in Descartes' terms, a tendency to motion.

Although Descartes had taken the trouble to examine a bull's eyeball, he usually had little time for experimentation, preferring the Greek approach of dealing purely with thought. Even so, he developed a solution to an oddity of light that had defeated many great minds from Ptolemy onwards – the bending of light as it passed from air to water or glass – refraction. While Alhazen had come up with a reasonable argument on *why* light bent, no one had successfully predicted *how much* light would bend by. Descartes decided that there was a fixed relationship between the angle at which light hits a substance and the angle at which it continues inside the substance (Figure 4.1), or rather the sines of these two angles.

'Sine' is a geometrical term. Think of a triangle with a right

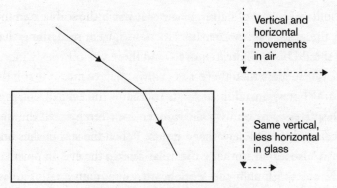

Figure 4.1 Descartes on refraction

angle in one corner. Choose an angle of the triangle, and the sine of that angle is the length of the side of the triangle opposite that angle, divided by the length of the longest side. So, for a 90° angle the sine is 1 (the longest side is always the one opposite the 90° angle), and the sine gets smaller and smaller as the angle does. Descartes correctly decided that the sine of the angle at which the light hits a piece of glass is always in the same proportion to the sine of the angle at which the light continues in the glass. It's quite amazing that he came up with these results at all, because his argument depended on assuming that light had a different speed in glass than it did in air.

Considering he thought that light didn't move at all, Descartes should have had a real problem here. Somehow he persuaded himself that, though light didn't move, it was a 'tendency to motion' and so could be treated as if it did. In practice he got the speed element of his equation the wrong way up, as he incorrectly assumed that light 'tended to move' faster in glass than air, but he will otherwise correct. This assumption by Descartes that light moved faster in glass than air wasn't quite as strange as it seems. Because of his view of what light was, it could only exist in matter, whether it was his invisible plenum or the glass. The more matter there was, he argued, the easier it should be for light to move – and there was obviously more matter in glass than there was in air.

What is amazing, though, is that without experiment, Descartes came up with the right rule for how much bending occurred. It may have been a pure coincidence, but another man also came up with the relationship at the same time. He was the Dutchman Willebrord van Roijen Snell. It may be unkind to suggest that Descartes took Snell's results and plugged them into his own shaky arguments (Snell, unlike Descartes, had got the ratio of speeds between glass and air the right way round), but it is interesting that this rule for the angle of bending is now called Snell's Law not Descartes' Law. Very little is known about Snell, who was professor of Mathematics at Leiden, or about his work, which has mostly been lost, but it is known that Snell undertook many experiments before coming to his result.

Fermat and the Baywatch principle

Refraction was now nearly as well understood as reflection. To make the final step of providing a theory that predicted the action of both took one of the greatest mathematical minds – and egotists – the world has known. Pierre de Fermat has had his spell of fame thanks to the drama surrounding 'Fermat's last theorem'. This seemingly simple mathematical proposition was finally cracked by the English mathematician Andrew Wiles in 1993. The actual problem dated back to the ancient Greeks, but Fermat, in a typically off-hand style, made it a challenge that would intrigue and infuriate generations of mathematicians.

The specifics of the problem could interest no one but a maths buff, but Fermat ensured (and subsequent books and TV series have proved) that the chase itself would be fascinating. Fermat loved to tease others with his wisdom by setting challenges. In a copy of the Greek writer Diophantus's *Arithmetica*, he added a note of his own, stating that while it is possible for the square of one number to be the sum of two other squares, a similar relationship isn't possible for cubes or any higher types of numbers. He then threw in a single line of Latin that was to set off the whole Fermat industry:

Cuius rei demonstrationem mirabilem sane detexi hanc marginalis exiguitas rum caperet – I have a marvellous demonstration of this proposition which this margin is too narrow to hold

Whether or not 'Fermat's last theorem' was just a boast or a joke on the part of the great man will never be known. Certainly the proof that was produced in 1993 would have been as incomprehensible to the seventeenth century mathematician as it is to any lay person now. But Fermat had shown once and for all his ability to tantalize. Luckily he was more forthcoming on the subject of refraction, pulling it into line with reflection by using a single, very different approach. In keeping with his personality, his interest in refraction seems mostly have been to spite Descartes, to whom Fermat took a powerful dislike. He felt that Descartes was much too imprecise, that he was inconsistent (saying that light had no measurable speed, but then 'proving' how refraction worked by a method that compared speeds) and that he made his deductions from analogies rather than rigorous proofs.

Before looking at Fermat's result it's worth thinking for a moment about the way he went about it. So often great breakthroughs have come by looking at a problem in a totally different way. Such an approach may require no new information, but suddenly the problem is transformed. The technique that Fermat used is a singularly powerful one for exploring the workings of the world. It's exactly the same technique that Richard Feynman would use to explain the fundamental nature of light many years later. It is called the principle of least action or the principle of least time, but what it amounts to is that nature is lazy.

In the world of solid objects, the principle describes why a basketball follows a particular route through space on its way to the basket. It rises and falls along the path that keeps the total difference between the ball's kinetic energy (the energy that makes it move) and potential energy (the energy that gravity gives it by pulling it downwards) along the whole of its journey to a minimum. Kinetic energy increases as the ball goes faster and decreases as it slows. Potential energy goes up as the ball gets higher in the air and reduces as it falls. The principle of least action establishes a logical balance between the two.

This principle can also be applied to the way light behaves. The whole business of refraction seems odd to begin with. Light is travelling happily along in a straight line through the air. It hits a piece of glass at an angle. Suddenly, for no obvious reason, it changes direction, bending down into the glass instead of carrying on in a straight line. This doesn't make a lot of sense until you apply the time version of the principle. The principle of least time says that light gets to where it's going as quickly as possible. Fermat had to make two assumptions to apply this – that light's speed isn't infinite (the speed of light was yet to be measured in 1661 when Fermat produced this result), and that light moves slower in a dense material like glass or water than it does in air.

We are used to straight lines being the quickest route between any two points – but that assumes that everything remains the same on the journey. In this case, light was travelling faster in air than it was in glass. Because of this, a straight line was no longer the quickest route. To see why this is the case, compare the light's journey to that of a lifeguard, rescuing someone drowning in the sea. The obvious route is to head straight for the drowning person. But the lifeguard can run significantly faster on the beach than she can run or swim in the water. By heading slightly away from the victim, taking a longer path on the sand, then bending inwards and taking a shorter path in the water, the lifeguard can get there quicker (Figure 4.2). (This analogy has led to Fermat's approach sometimes being called the *Baywatch* principle.)

Figure 4.2 The Baywatch principle: the solid line is quickest

In just the same way, a light ray could get from its start point in air to its end point in the glass by making a straight-line journey. Or it could be angled a bit closer to the horizontal as it travelled through the air, but then bend when it hits the glass so it still reaches the same end point. Because of the change in angle it will have a longer journey through air and a shorter one through glass. And because light moves faster in air, it will take less time to follow the bent route than the straight line. But bend the light too much and it has to travel too far in the air to overcome the advantage of less time in glass. The angle that minimizes the journey time is exactly the one that actually occurs.

The strange case of Iceland spar

With Fermat's masterly application of mathematics to the problem of refraction that had troubled observers for 2000 years, it seemed as if light had given up all its important secrets. True, no one knew exactly what light was, but its behaviour was predictable. There was one less problem for science to concern itself with. But light cannot be pinned down so easily. In 1669, Erasmus Bartholin, an enthusiastic experimenter from a Scandinavian family that seemed to specialize in breeding medics and scientists, published the snappily titled *Experimentia Crystalli Islandici Disdiaclastici*, believing he had made a breakthrough. There was, thought Bartholin, not one type of light, but two, identical in appearance but differing in behaviour.

It is a challenging thought. The natural inclination is to apply the decisive blade of Ockham's razor. William of Ockham, a contemporary of Bacon known as Doctor Invincibilis (the invincible doctor), devised this remarkably powerful principle, saying that entities should not be multiplied unnecessarily – in plainer English, that we should accept the simplest realistic assumption. The sensible starting point, then, is that there is only one entity we call light. It may come in different colours, be visible and invisible, but it's all the same, basic phenomenon. Yet Ockham's razor is a convenience, not an unchallengeable route to the truth, and Bartholin had observation on his side.

It was all a matter of the strange behaviour of Iceland spar. Bartholin was fascinated by crystals, particularly this clear form of calcite, which grows in small slabs shaped like a distorted brick. Calcite is a common enough mineral, the natural form of calcium carbonate, second only to quartz in quantity on the Earth. It is the main constituent of limestone and marble. It supports the complex structures of seashells. Yet despite being part of such a common family, the clear crystalline form of calcite brought Bartholin to his strange conclusion.

When he put a block of Iceland spar on top of a piece of paper with a straight line drawn on it, it was no surprise to Bartholin that the line should be moved a little out of place by refraction. But his unexpected discovery was that he saw not one but two lines. It was as if there were indeed two kinds of light, one bent more than the other by the crystal. Unlike the observations Newton was making at around the same time with a prism, there was no splitting into colours by this flat block, just two, clear separated images.

The truth behind this mystery would remain impenetrable for two hundred years, but Bartholin's concept of two types of light was a first attempt to explain it. Even so, the phenomenon may have been put to practical use long before. The Vikings possessed a special gem known as a sunstone. While no details of the sunstone have ever been uncovered, it has been suggested that this might have been Iceland spar, and could have been used as a navigation aid to take a sight on the Sun when it is hidden behind clouds.

What is certain is that clear crystalline calcite proved valuable during the Second World War. The degree to which the second image is shifted depends on the distance away of the object seen through the crystal. Iceland spar was used in bombsights, crude instruments carried on bomber aircraft to estimate the distance to the target. Even today, Iceland spar is used in specialized optical instruments because of the effectiveness of the image separation.

Iceland spar's secrets would be probed by Newton's contemporary Huygens and finally explained by Victorian scientists. Yet its strange behaviour was eclipsed by a venomous debate that shook the scientific world. A war of words was about to break out over the nature of light and colour.

Chapter Five

Seeing further

Nature, and Nature's laws lay hid in night.
God said, Let Newton be! and all was light.
ALEXANDER POPE – EPITAPH: INTENDED
FOR SIR ISAAC NEWTON

The telescope and the microscope opened windows into the physical world, but left untouched the mystery that made them possible – light itself. It was left to a young Englishman, trapped at home for two years by the plague, to brush aside 3,000 years of philosophical fog and precipitate one of the fiercest disputes science would ever see.

Isaac Newton's home was an unlikely cradle for scientific genius, a fact that is obvious despite the romanticized glow that obscures much of his life. No other scientist has suffered such embroidery of the facts. Only Einstein received the same degree of adulation, but under the more discriminating spotlight of modern journalism. Newton's life story was rewritten as assiduously as that of any movie star in the golden age; to see the truth we have to sift through highly rose-tinted histories.

Gentleman farmer

Newton's upbringing was more appropriate for a farmer than a professor, yet the personality traits of independence and determination that stuck with him for the whole of a long life were formed in those early years. His home, technically a manor house, was little more than a large stone-built farm in the Lincolnshire village of Woolsthorpe. The indeterminate status of the building perfectly reflects the Newton family's social standing. For four generations they had clung to the lower fringes of genteel respectability before Newton's father, Isaac senior, reinforced his position by marrying Hannah Ayscough.

The marriage was helpful for both families. The Ayscoughs' higher social standing gave Newton's father a boost in Lincolnshire society. But lack of cash had made life difficult for the Ayscough family and Isaac senior's relative wealth was an attractive proposition. If it hadn't been for the Ayscoughs, with a history of sons being sent to university (in those days this meant Oxford or Cambridge), it may well have been that young Isaac Newton would have spent his days around Woolsthorpe, living the life of a bored gentleman farmer. Whether this was what his father intended for him will never be known. His parents were married in April 1642; before the year was out, Newton's father was dead.

Exactly what happened to old Isaac has not been recorded. Newton himself tried to cloud the details of this period by giving a false date for his parents' wedding when he was knighted, eager to cover up any suggestion that he had been conceived illegitimately. As it was, he was born on Christmas morning 1642, premature, small and frail. For three years he and his mother lived at Woolsthorpe alone but for the servants. She had few financial worries, and the infant Newton enjoyed a pleasant if quiet life around the comfortable farmhouse until his mother's second wedding in January 1646. The marriage was to have a dramatic effect on the young boy.

Hannah's new husband, Barnabas Smith, was rector of the nearby village of North Witham. The role seems to have been of little importance to him – such positions in the seventeenth century Church of England were often sinecures, requiring a minimal interest in the parish and its parishioners. To be an absentee rector wouldn't have raised any eyebrows, but despite this it was Hannah, not Smith, that moved house after the wedding, making the short journey to North Witham. Young Isaac did not. He was left behind at Woolsthorpe with Hannah's parents, who took over the manor house to look after him.

This behaviour seems strange today, casting Smith as an archetypal evil stepfather, but at the time it was a thoroughly reasonable move. Smith was older than Hannah by at least thirty years. He hoped to have his own children – in fact he and Hannah were to have three before his death in 1653. Isaac would simply have been in Smith's way. It's easy with post-Freudian hindsight to ascribe Newton's later solitary lifestyle to this difficult separation, and to the noisy imposition of his unwanted half brother and sisters when his mother returned after Smith's death. Certainly there is evidence from the scraps of writing left behind from Newton's teens that he felt no love for his stepfather and the new regime. He was deserted, left behind with elderly carers who could not give him the emotional support he needed. The result was a fierce independence that earned Newton a lifetime reputation for aloofness.

Discovering education

It was probably with some relief all round that in 1654 he was sent to study at the King's School in Grantham. A mere seven miles from home, this was still too far to travel daily at a time when transport was difficult in rural areas, so Newton was found a place with a respectable Grantham family, the Clarks. While young Isaac certainly benefited from the experience of a structured education – and the headmaster of King's School, Henry Stokes, was to play a significant part in getting Newton to Cambridge – it was the head of the Clark family who was most responsible for exciting Newton's interest in science.

Mr Clark was an apothecary, the equivalent of a modern pharmacist, but with a much greater freedom to experiment and prescribe his own remedies. His shop was a treasure trove of brightly coloured flasks, ranked above rows of polished wooden drawers, each hand-labelled with an exotic name suggestive of strange places and possibilities. Newton snatched at the opportunity to observe Clark at work and to help out. It was here in the shop that Newton's profound belief in the value of experiment was born. As he discovered the attractive certainty of science when compared with the unpredictability of human relations, Newton began to shine in his schoolwork. By the time he was 16, Newton's academic performance was such that Henry Stokes was preparing to enter him for university. At this crucial time in Newton's development, his mother decided to remove him from the school.

Hannah was determined that Newton should put the farm first. It was both his right and his responsibility in her eyes. But Newton had no intention of complying with his mother's wishes. He escaped from the mundane farm work at every opportunity. When a servant was given the job of keeping him busy, Newton instead got the servant to do the work for him. He was always taking the opportunity to read, to gather knowledge. Even when sent to market in Grantham he spent most of his time in Clark's shop, once more leaving the servant to handle the business. Henry Stokes heard of Newton's determination to study and asked Hannah to reconsider. Supported by Newton's grandfather, a fellow Cambridge graduate, Stokes finally persuaded Hannah that allowing Newton to apply for university made practical sense. In June 1661, Isaac Newton went up to Trinity College, Cambridge.

Even after Newton's move to Cambridge, his mother's distaste for his escape from the farm showed through. She only provided him with an allowance of £10 a year on top of his academic fees. This £10 has to be put in the context of a typical labourer's earnings of around £4, but it amounted to less than a week's income for his mother. Because of this lack of financial support, Newton entered the university as a sizar, a position that required him to act as a servant to support himself, rather than taking the expected more costly and comfortable position of pensioner.

At the time there were few subject options, with a curriculum that was heavily influenced by the Greeks. But students were at least allowed to cover philosophy, a catch-all that included the work of more modern scientists like Galileo. It is obvious from Newton's notebooks that from the very beginning he was prepared to question what he was taught, assembling his own ideas instead of consuming accepted wisdom whole. He had already rejected the conventional theory of the four elements of earth, air, fire and water devised by Empedocles, preferring the atomist view. This stated that everything is composed of minute indestructible particles, and would heavily influence Newton's thinking throughout his career. By 1664, his urge to go his own way brought him to experiment with light.

The Stourbridge prism

It started with a fair. Stourbridge Common, on the side of the river Cam between the villages of Chesterton and Fen Ditton, is now solidly embedded in the sprawl of Cambridge, but in Newton's time it was sufficiently far from the city to be outside the bounds of the university. University members were monitored by a private police force, the proctors, who attempted to prevent them from drinking in taverns or mixing with tradesmen. Sited outside the proctors' influence, the annual Stourbridge fair was an opportunity for university members and townsfolk alike to let their hair down. On the long, thin stretch of common was an intriguing array of entertainments and refreshments, with abundant stalls selling unusual charms and toys. It was at the 1664 Stourbridge fair that Newton bought a prism.

A simple block of glass with a triangular cross-section, like a Toblerone box, the prism's ability to produce a miniature rainbow, just as the water drops of rain did, was already well known. Newton took his new toy back to his rooms and managed to display a faint spectrum of colours from the light leaking through a hole in his blinds. This pale strip of light inspired him to take a dramatic leap of imagination. The popular theory of the time was that the shape of the glass in the prism tinted the pure white light in different ways. Newton was sure that the different colours making up the spectrum were already contained within the whiteness and were merely separated by the prism. This was the first real example of Newton firing on all cylinders, developing a totally new proposition to replace the accepted view.

He got hold of another prism and allowed the rainbow from the first prism to fall on the second. If the prism really was tinting the light as everyone said, then it seemed reasonable that different hues should be produced when the coloured light itself passed through another prism. Instead the colours remained unchanged. Newton felt triumphant. He had met a deep-seated need to prove his ideas by experiment. Now he could build on his theory to deduce correctly the mechanism by which an object appears to have a particular colour. If an object is struck by white light and absorbs some of the colours, then it will return the remaining mix of colours that weren't absorbed, producing the apparent colour of the object. For instance, an apple that absorbs the spectrum from red to yellow at one end and blue to violet at the other will be left returning green light to the eye – the apple will look green.

Newton also began to question the very nature of light itself. Although Empedocles' idea of light emanating from the eyes had been excluded from the ancient philosophy that was still taught, Newton was not comfortable with Descartes' modern theory. Newton, convinced of the existence of atoms, believed that light, like matter, was made up of small particles – corpuscles in his terminology. He was dismissive of poor Descartes. Newton pointed out that we should be able to see in the dark if light were a matter of pressure as Descartes suggested – running along would be enough to put pressure on the eyes, producing sight. Newton's distaste for Descartes' theory had ample opportunity to grow in the thinking time afforded him by the plague.

Trapped by the plague

Disease was no stranger to the seventeenth century, but by the mid-1660s, it seemed that public health was steadily improving.

London had gone 15 years without a major infection, so when an outbreak of bubonic plague struck the capital in the severely cold winter of 1664 it took the population by surprise. It was only the outbreak of fire in Pudding Lane, cauterising a fair proportion of the city, that tamed the Great Plague. London was not the only place to suffer. Cambridge had its own outbreak, forcing Newton to spend two years at home in Lincolnshire.

Newton was 23, recently graduated with a less-than-exciting second class Bachelor of Arts degree. With two years enforced leave from study, most of Newton's contemporaries put their academic careers aside and turned to pleasurable pursuits. Little more would have been expected of Newton than to stroll round the family farm, keeping an eye on the harvest. Instead, in that period of two years, Newton is said to have produced a lifetime's work. As legend has it, he grasped the workings of gravity, spurred on by the sight of an apple falling from a tree in the orchard just outside the manor house. He developed calculus, the mathematics of change that was necessary for every scientific development of the twentieth century. He explained the movements of the planets. And he produced his theory of optics, light and colour.

If true, it was a superhuman accomplishment. In 24 months of concentrated effort Newton had set the scientific agenda for the next two hundred years. The doubt isn't that Newton achieved all these things – but rather how much of the discovery was packed into those two years. It's certainly the case that he laid the groundwork in each subject, and that the opportunity to think with little disturbance would have been of great value. Yet it is more likely that much of the substance that the Newton myth ascribes to those two years was developed later on. Even so, of all the possible developments during that time, the most certain is his progress in the study of light.

That he was able to see at all, let alone study, when he reached Woolsthorpe was more due to luck than good sense. Before leaving Cambridge, Newton had been experimenting with his own eyes' response to light. His attempts could easily have left him blind. He first stared into the Sun's reflection in a mirror, repeatedly turning his eyes away to a dark corner of the room to observe the spots and colours that floated in the darkness as an after image. This left him unable to see at all for several days. As if determined to ruin his sight, he then experimented with the effect of the eye's shape on vision by inserting a thin knife between his eye and its socket and putting pressure on his eyeball. That light was something of an obsession for Newton cannot be doubted.

Experimentum crucis

Once he returned to Cambridge from the country, Newton began to gain honours with startling rapidity considering his uninspiring degree result. It must have been that the two years of isolation had allowed him to mature into a much more attractive prospect. He became a fellow of Trinity College in 1667, gained his MA in 1668, and the next year, while still only 26, became the second holder of the Lucasian Professorship in Mathematics, a position later held by Stephen Hawking. This new job obliged Newton to give occasional lectures, the first of which was on light.

It would be exaggerating to say that Newton's lectures were popular with the students. In fact, after the first, many were entirely unattended. Newton would trim his delivery by half from the expected thirty minutes, but still resolutely insisted on speaking to an empty, echoing room. With the greater freedom of his professorship, Newton was also able to spend much more time on experiment. To isolate a single colour (or at least what the eye sees as a colour – a spectrum in fact consists of an innumerable range of colours, each blending into the next), he put a card with a hole in it next to a prism, only letting through a narrow band of light. Not only did he confirm his view that when this beam was passed through a second prism no different colours were produced – red light remained red, blue remained blue and so on – he discovered that red coloured light was bent much less by the prism than blue light.

The degree of bending, the refraction, varied as he moved through the different colours. He later referred to this discovery as the *experimentum crucis*, the crucial experiment, emphasising its significance as a turning point in the understanding of the nature of light. He had found something fundamental and new – that light was made up of colours that were distinct entities, impossible to change one into the other, each bent differently by a prism. For good measure, his experiment explained why a prism worked at all.

When a beam of light hit an ordinary block of glass there was no rainbow produced. As the light passed from air to glass it was true that the blue light would bend further than the red, splitting it out, but when it reached the far side of the block it would move back the other way an equal amount and the result would be to recombine the colours. The prism's triangular faces meant that the two opportunities to bend – towards the vertical of the first face and away from the vertical of the second – both resulted in movement in the same direction. The colours remained separate.

It's hard to imagine now just how dramatic Newton's ideas

were at the time. To even the greatest seventeenth century thinkers, the concept at the heart of Newton's theory seemed utterly strange. Newton was saying that white light was made up of an infinite range of different colours, *impossible to change into each other*, each bending differently when they passed through a prism. Yet everyone knew that you could easily make one colour from two others – a child could do it with a paint box. So it seemed ludicrous to suggest that colours were absolutes that could be temporarily mixed in white light but would always retain their separate identity. There was also rampant confusion between the colour of light and the colour of objects, which Newton had to go to great lengths to explain.

Perhaps even more than this confusion, Newton was fighting the ghosts of ancient Greece. In the university of his day, the authority of the Greeks was still held in high regard. Their linkage of colours with perception (and Aristotle's highly impractical idea that all colours were produced by mixing black and white) were hard to shift in the conservative world of academia.

It shouldn't be too surprising that our current view of colour is quite modern – the division of the rainbow into red, orange, yellow, green, blue, indigo and violet is, after all, entirely arbitrary. The rainbow colours form a continuous meld from one end of the spectrum to the other. (The actual spectrum produced by the Sun has some black lines due to the absorption of certain colours by the material in the Sun, but in principle the spectrum is continuous.) The arbitrariness of what we now consider to be 'obvious' colours is clear from the fact that there was no word for orange in Europe until the tenth century, and then, until the 1600s, that colour was only associated with the fruit from which the colour takes its name.

To make it even clearer that white light was composed of

the spectrum's colours, Newton used a lens to recombine a spectrum to a blob of white – and even showed that this mixing was reversible by using a fine comb to let through only certain parts of the light before it was focused, producing a different coloured blob, the colour depending on which parts of the spectrum he let through. Newton's understanding of the ways different colours refracted was to be invaluable when, soon after, he constructed a reflecting telescope. The curved mirrors of his telescope did not suffer from the distorting coloured fringes – chromatic aberration – that plagued the lens-based telescopes of the time, which were bending different coloured components of light to a different degree.

Battling Hooke

It was Newton's telescope, rather than his theoretical work on light, that drew him to the attention of the Royal Society, where he was elected a fellow in 1672. The Royal Society was a talking shop on natural philosophy, founded 24 years before. As happy to discuss the vivisection of crocodiles and the existence of werewolves as the mechanisms of light, the Society's members had a common fascination with the workings of nature. In Newton's day the Society met in Arundel House in London, and had already become the foremost testing ground for new theories in the sciences. Newton's success with the telescope and a growing correspondence with the Society's secretary, Henry Oldenburg, led him to send Oldenburg a letter detailing his theories on light and colour that was quickly published in the *Transactions of the Royal Society*.

From their correspondence it is obvious that Newton got on very well with Oldenburg. The same could not be said for every member of the society. In particular, Newton found an uncomfortable foil in the society's curator of experiments and later secretary, Robert Hooke. Nothing at Cambridge had prepared Newton for such aggressive opposition, but it is here that his upbringing showed through. When Hooke attacked his theories, Newton was quick to respond.

Hooke was an impressive adversary. A few years older than Newton, he had a much broader range of interests and an easy way with the social in-crowd that Newton couldn't match. Where Newton's life was confined to the near-monastic isolation of his rooms at Cambridge, Hooke was a popular member of the coffee house set and a reputed womanizer. Hooke thought Newton self-centred and inept; Newton considered Hooke a shallow dandy.

In his role as curator, Hooke had to study and comment on Newton's paper. Since he disagreed with Newton's hypothesis that light was made up of particles (which had nothing at all to do with the paper on colours), Hooke practically ignored what Newton had written. He later admitted to spending no more than three hours on it. His review made it seem that Newton had no sound basis for his arguments on the nature of colour, yet these arguments owed everything to experiment and nothing to Newton's more speculative theories.

Newton struck back, suggesting that Hooke had not properly understood his reasoning. This pressed Hooke into responding by turning Newton's own paper against him. Unable to find a scientific error in Newton's work, Hooke instead tried to trip Newton up on procedure. At the start of the paper, Newton makes a plea for clearly separating fact and hypothesis. Later on, after describing his experiment with the prism and the lens, Newton makes the statement:

These things being so it can no longer be disputed
whether there be colours in the dark, nor whether
they be the qualities of the objects we see, nor perhaps
whether light be a body.

Hooke pounced on this last remark. By suggesting that
light was a body, Newton was insinuating his hypothesis that
light was made up of particles into his experimental statements
of fact about colour, apparently contradicting his own philosophy. Hooke wrote to Newton, pointing out his error.

At this point, Newton's reputation was in the balance.
While he was hidden away at Cambridge considering his
response, Hooke took the opportunity of his position to press
the attack – his feelings about Newton had quickly become
more personal than professional. Hooke informed the Royal
Society that *he* had invented the reflecting telescope before
Newton, indirectly accusing his country rival of stealing ideas.
Luckily for Newton, this wasn't the first time Hooke had made
claims he couldn't back up. He was in the habit of boasting of
achievements he had made, without being quite able to produce
the results. The members of the Royal Society had seen it all
before and weren't impressed.

This didn't stop Hooke looking for opportunities to do
Newton down, and before long the still-young and guile-
less Newton's rampant enthusiasm gave Hooke a lever to use
against him. Newton had received through the Society a series
of letters from a Parisian Jesuit priest and professor, Ignance
Gaston Pardies. A supporter of Descartes, Pardies dismissed
Newton's idea that white light was composed of the many colours of the spectrum. Newton was not subtle in his replies,
suggesting that Pardies was an amateur. He instructed the

priest in a heavy-handed way, as if he were dealing with a village idiot rather than a respected scientist. Hooke complained to the Society about the tone of the published exchange of letters, which earned Newton a mild rebuke from Henry Oldenburg.

Newton was furious. In a fit of pique he decided to keep back his newest theories on light that would much later form the basis of his book, *Opticks*. This was the first of two setbacks for publication of the book, which would now not see the light of day until 1704, more than 30 years later, when Newton was already an old man. The second delaying factor was not an argument, but a fire. You can still see Newton's old rooms, number four on staircase E in the Great Court of Cambridge's Trinity College, but there is no trace of the wooden outbuilding that was crudely nailed on to the grand structure of the court to house his experiments. Five years after Hooke's intervention, a fire broke out in this makeshift laboratory while Newton was attending a service in Trinity's starkly classical chapel. The fire was probably caused by Newton's experiments on the borders between chemistry and the then still-respectable alchemy. It destroyed many papers, including a manuscript that was probably the first draft of *Opticks*.

Although Newton held back his book, he was less restrained about tearing into Hooke. Once he had assembled his defence he proceeded to pull Hooke's arguments apart, destroying his complaints line by line in a logical *tour de force*. Hooke was ordered to re-think his assessment of Newton's original paper. It seemed that Newton's views would now triumph. But though Hooke was still to be a severe irritation to Newton, this London dandy was not the biggest thorn in Newton's side. The real intellectual threat was from a man for whom Newton had much more respect – the Dutch scientist Christiaan Huygens.

A new challenger

Huygens had given a very positive response to Newton's original paper on the origin of colours, though from the wording
of his letter he totally misunderstood it. Now he came onto the
attack, echoing Hooke's objections in dismissing not Newton's
discoveries but his methods of reasoning. Huygens used much
more rigorous argument than Hooke and displayed none of the
other's evident bad feeling. Unlike Hooke's barbed missives,
Huygens' letter was perfectly discreet and polite. But by now
the always-touchy Newton was on a hair trigger. He was frustrated that Huygens, Pardies and Hooke had all picked up on
his purely speculative remark about light possibly being made
of particles, using it to argue against his quite separate observations of the way white light was composed of many colours.
His response was to tender his resignation to the Royal Society,
claiming that it was inconveniently distant from Cambridge,
but in reality because it seemed to have become a channel for
others to attack him.

The unflappable Henry Oldenburg proved himself a better
tactician than Newton by calling Newton's bluff. He calmly
offered to cancel Newton's subscription to the Society. This
unemotional response seemed to deflate Newton's anger. He
took the threat no further and soon after replied to Huygens
with a relatively moderate letter suggesting that the Dutchman
should duplicate the prism experiments before disagreeing with
the results. Huygens had no interest in a fight and so the matter subsided. Newton had three peaceful years in Cambridge
before putting his head over the parapet again – and once more,
Hooke was waiting for him.

Newton submitted two papers to the Royal Society, one

explaining how the reflection, refraction and diffusion of light could be produced by the action of small particles, and a second detailing a series of experiments that he hoped supported this theory. (It was typical of Newton's approach to split his observations into hypotheses, experiments and the logical deductions that could be made from those experiments.) Hooke, probably quite genuinely, felt that Newton was stealing ideas from his book *Micrographia*. Some of the conclusions were similar, even though they had started out from totally opposing theories.

With his fingers burned by his failed attempt to disgrace Newton through the Royal Society, Hooke began to use the environment where he had a natural advantage over his country rival – the coffee houses. In discussions with his friends and cronies he ensured that his opinions on Newton's lack of originality became widely known, and he encouraged his listeners to spread the word widely. With rumours beginning to fly about Newton's dishonesty, Hooke's obvious vehicle for finishing off his rival was to use the Royal Society once more. But Hooke's relationship with Henry Oldenburg was going downhill fast. Instead of putting his case through the formal channels of the Society, Hooke began to write to Newton directly.

This was a bizarre decision. Not only did the use of private letters mean that the debate would not receive the oxygen of publicity, it forced Hooke and Newton in his turn into the artificially polite environment of seventeenth century personal letter writing. It was as if Hooke had challenged Newton to a duel, then selected feather pillows as the weapon. The insults and arguments were still there, but all sharpness was lost in the stilted surroundings of pleasantry. Take Hooke's assertion that Newton had stolen ideas from *Micrographia*. The need to be excruciatingly polite took away all the intended venom:

I [...] am extremely well pleased to see those notions
promoted and improved which I long since began, but
had not time to complete.

Newton came back with a study in formal compliment and
respect. His letter contained what has become the single most
quoted phrase in the history of science. It is strange that it
should appear here, in the midst of a painfully polite series
of arguments and insults. In fact, the remark itself has been
interpreted as an insult to Hooke, who had a deformed back
that made him seem particularly small in stature. Hooke was
very sensitive about his appearance, never allowing his portrait
to be drawn or painted. It's easy to feel the barb of sarcasm
when Newton remarked after commenting on Descartes and
Hooke's work:

If I have seen further it is by standing on the shoulders
of Giants.

No one could consider Hooke a giant.

Leaving Cambridge

As the squabbles with Hooke petered out, Newton got back to
the safer world of natural science, moving on from light to the
study of motion and gravity, taking the steps that would result
in his masterpiece, the *Principia*. Before he was to think much
about light again, the fame of this book and the strength of his
faith would carry the once-retiring scientist into the political
limelight.

Newton became embroiled in politics when the Catholic

King James II vowed to force Catholic students on the resolutely Protestant university. Newton was part of the group that attempted to resist the king, though they were to crumble when the king sent the notoriously merciless Judge Jeffreys to deal with them. But time was on Newton's side. Within a few years William of Orange had brought a Protestant monarchy back to England and the threat of a Catholic invasion of Cambridge had disappeared.

It seems, though, that Newton gained a taste for politics, or at least the life of a politician. In 1690 he became the MP for Cambridge University, a post he held for a year, during which the only speech he is known to have made was asking an usher to close a window because he felt a draught. His political dabblings and increasing fame brought Newton into a much smarter circle, and that meant big changes in his personal circumstances.

For a short while it seemed as if this disruption of his secluded life was going to throw him into madness. During a few months in 1693, he began writing strange letters to friends and associates. Diarist and Royal Society president Samuel Pepys was informed that Newton could never see him again due to some unspecified disgrace, while the philosopher John Locke was accused of attempting to embroil Newton with women. But soon Newton had recovered, his position of high favour continued, and it was no great surprise to his associates that in 1696 he abandoned Cambridge to take on the high society position of Warden of the Royal Mint. He never returned to a pure academic life.

Newton's predecessors in the number two job at the Mint had seen the post of Warden as an honorary position, but Newton was put there for his practical skills. He proceeded to shock the rest of the seventeenth century management team by his diligence, often turning up as early as four in the morning and staying late into the night. His energies were needed – Britain's coinage was in a dire state and required wholesale replacement.

Newton's supreme contribution was to optimize the processes of the Mint, proving as effective a management consultant as he had been a scientist. He threw himself into the task and the accompanying role of criminal investigator, winkling out counterfeiters and clippers (who sold precious metals, melted down from slivers of the edges of coins). His pursuit of criminals was merciless, despite the extreme punishments of the time. The worst of those he pursued were hanged until nearly unconscious, their bowels cut from their living bodies and then the remains sliced into four quarters. Newton was unmoved by their fate; his worldview did not admit compromise.

At the turn of the century he stepped into the shoes of his dead superior as Master of the Mint and went on to undertake another uneventful stint for Cambridge University in parliament. Yet the challenge had gone from his business career and another death revitalized Newton's scientific fervour. In 1703, Robert Hooke passed away. In the following 12 months, Newton was re-elected to the then failing Royal Society, rapidly took over its presidency, and then turned around the Society's fortunes with a combination of prestige and organizing ability. A year later came the publication of the *Opticks*.

This three-volume book puts forward his original theories of light and colour in more detail. It is still surprisingly readable – Newton resists the urge to slip into jargon and avoids the bland indirectness of modern textbooks. There are also many 'queries' – unanswered questions and untested hypotheses. In fact, Newton's original intention was to have a fourth book. Convinced that light truly was composed of small particles, Newton hoped to pull together his elegant theories of motion and gravitation with his observations on light to form what would now be called a Grand Unified Theory or a Theory Of Everything. This idea of combining the different forces and fundamental phenomena of creation into a single theory has since obsessed many of the great minds of the twentieth century, from Einstein to Hawking. Like Newton, they were not to succeed.

Although what Newton was attempting was less grandiose than the schemes of his twentieth century counterparts, it still went far beyond what was practical with the science of the day. Rather than publish a whole book of speculation, he incorporated the queries into the third volume, giving pointers for the future. As always, he was not happy to tread the path of many of his contemporaries in putting forward complete theories based on hypothesis – he wanted his exploration of light to be solidly grounded in experimental observation. As there was no way to achieve this for his corpuscles of light, or his attempts at unifying light and his studies of matter and motion, they remained outside the main body of the work. Some of his queries were way off beam – others had an eerie closeness to discoveries that would not occur for hundreds of years.

In Query 1, for example, he wondered if bodies act upon light at a distance, bending its rays. This fits with his attempt to pull gravitation and his model of light as microscopic particles together, hoping that light would behave just as the planets do in being pulled out of their natural straight lines courses by the Sun. But it wasn't until Einstein's general theory of relativity (see page 208) that such a possibility was once more proposed. In

May 1919, photographing a solar eclipse off the West African coast, English scientist Sir Arthur Eddington was to prove that Einstein's theory and Newton's query were correct.

Newton lived until 1727, gaining a knighthood and continuing to sustain a hold on society that seemed designed to disprove the doubtful possibilities of his birth. Just as he had moved from experimental science to business consultancy, he brought a powerful energy to his political manoeuvring as President of the Royal Society, still as ready to rise to a challenge as he had been with Hooke, but now acting from a position of formidable power.

In his battles with Astronomer Royal John Flamsteed (a onetime associate of the hated Hooke), and even more so with the German mathematical genius Gottfried Wilhelm von Leibniz, Newton was vitriolic. He managed to do serious damage to Flamsteed's career, but in Leibniz, Newton found his match both in intellect and force of character. While each independently had devised the mathematics of change, calculus, in the 1660s and 70s, Newton (who called it the method of fluxions) found that his claim to be the sole originator would not be upheld. For once, despite a forty-year campaign to prove Leibniz a cheat, he was forced to lose an argument, though he did so with little grace.

Newton was not to publish anything more of significance, though *Opticks* went through a number of revisions (strangely, as it now seems, being translated from the original English into a version that, like the *Principia*, was written in Latin). The later editions contain a number of further intriguing queries, but no new experiments were added. Newton was justifiably able to rest on his scientific laurels, continuing with his politicking until he finally died at the age of 85.

Looking back on Newton's life it is fascinating how many parallels there are with Galileo (see page 64), who died the year Newton was born. Both men fought against the traditional philosopher's approach of basing theories on pure reasoning, preferring to deduce from the realities of experiment. Galileo nurtured the seeds of modern scientific thinking that Roger Bacon had sown and planted; Newton harvested the crop. Both Newton and Galileo's primary claims to fame involved motion; both were fascinated by light; each has a type of telescope named after him. Both lived to a ripe old age, both were politically adept and associated with a scientific society in its early years. Their circumstances were very different, but each combined a fascination with natural science and the inability to let things go, a determination that shaped their lives.

Making waves

Although Newton's critics found it difficult to accept that white light was composed of the spectrum of colours, the bitterest arguments centred on his belief that a beam of light was made up of a stream of particles. This was unacceptable to Hooke, Pardies and particularly Huygens, who had a near-religious conviction that light was a type of wave.

Drop a stone into a still pond – you will see a series of ripples moving out from it in circles. These are the sort of waves Huygens imagined making up light. It was already thought that sound travelled in waves, and it seemed reasonable that light should as well. Newton had not been convinced. Anyone that thinks light is a wave has two big problems to overcome. The first is, what is it a wave *in*? This seemingly innocent question really is a problem. Think of the water, and what the wave consists of in that water. What we call a wave is actually the water moving up and down in a regular pattern away from the stone that was dropped in it. The wave transfers energy across the water like a chain of people passing boxes along – the people (the water) don't move sideways but the boxes (the energy) do. When we hear sound, it is movement of the air that allows the wave to pass along. But light travels across apparently empty space. What is rippling when light passes by?

To cope with this ability of light to cross the void, an invisible 'something' filling empty space, called the ether (in those days spelled 'aether'), the successor to Descartes' plenum, had to be invented. Light waves were assumed to be ripples in this ether. For Newton, though, it was the second problem of light as a wave that was the real stumbling block – the complex sounding (but actually very simple) matter of rectilinear propagation. Waves expand outwards. Think of water again. Send water waves through a slot in a piece of wood and they will open out beyond the slot. Sound does this too – we have no problem hearing round corners, the sound wave opens up around the blockage and reaches our ears. So why can't we see round corners? Send light through a slot and all you get round the corner is shadow. This doesn't seem normal behaviour for waves, but fits well with Newton's picture of a stream of particles.

Throughout the eighteenth century it didn't really matter whether Newton's particles or Huygens' waves were more sensible – the cult of Newton, the first scientific superstar, was such that his opinion was considered law in England and revered throughout much of Europe. This was rather unfortunate, as waves had a lot going for them. Even Newton's main argument about light not going round corners wasn't as much of a clincher as it seemed.

Back in 1665 a Jesuit priest, Father Francesco Grimaldi, published a little book on light. He described how the shadow of a round object held in front of a light was actually smaller than it should be, as if the light was creeping around the edges just as waves would – and the border between shadow and light was not clear and exact as you would expect it to be with Newton's particles. Instead it was fuzzy. Looked at closely, there seemed to be little fringes of light and dark surrounding the shadow. It didn't fit the picture of tiny particles moving in a straight line. Grimaldi's discovery was already in place to give weight to Huygens' case.

When Huygens described light, he started from a very similar point to Descartes. This is not surprising – Descartes was a friend of the family, who occasionally visited the Huygens' home in The Hague. Like Descartes, Huygens thought that space was filled with a multitude of tiny invisible spheres, and it was through these spheres, making up the ether, that light moved. The big difference in Huygens' version was assuming that these spheres would compress like rubber balls, rather than being totally rigid as Descartes thought. This meant that a push at one end wouldn't immediately result in pressure at the far end. Instead a wave would start moving through the tiny balls as each squashed and then sprang back to shape, just as a sound wave moved through air, until it reached its destination.

This picture of waves moving through a sea of tiny balls allowed Huygens to come up with his most significant contribution to the understanding of light and waves. He said in his masterwork on light, the *Traité de la Lumière*, presented to the French Academie Royale in 1679:

Each particle of matter in which a wave spreads ought

not to communicate its motion only to the next particle
that is in the straight line drawn from the luminous
point, but also imparts some of it to all the others that
touch it. [...] So it happens that around each particle
there is a wave of which that particle is the centre.

In Huygens' picture, a wave is not a single ripple, but is
built up of a lot of tiny wavelets, moving outwards in all directions from each point on the front of the original wave (Figure
5.1). Generally the ones going off in odd directions would cancel each other out – as one goes left, the other goes right, and
the result is no movement. But heading in the direction of the
wave, there is no cancellation and the motion continues along.

This picture helps explain refraction – why light bends as
it enters glass. Imagine a thick beam of light hitting a piece of
glass at an angle. The first wavelet to enter the glass will slow
down, but at the far side of the beam, some wavelets are still
in the air, still expanding at full speed. The result, like a line of

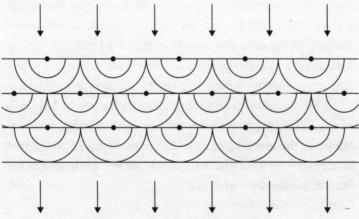

Figure 5.1 Huygens' wavelets in action

marching soldiers turning, will be to change the direction of the light, bending the beam into the glass.

Huygens' wavelets also show why the shadow of an object isn't quite as big as might be expected, as the tiny wavelets at the edge head in around the object – and because of that edge, there are no other wavelets heading in the opposite direction to cancel them out.

Adding the maths

Hugyens painted a very plausible picture of light as a wave, but there were still one or two gaps left in the maths. These were to be filled by a Swiss mathematician, Leonhard Euler, whose ideas come down to us with startling liveliness in the first real work of popular science, a series of letters to a German princess. Euler might have been born Swiss, but he lived a cosmopolitan life, crossing boundaries as easily as any of today's globetrotting academics. This intellectual high life was quite different from that envisaged for him by his father, a protestant priest with a limited income. It seemed natural to Euler senior that his son should follow him into the ministry, especially as Euler's mother also came from a family of clergymen.

As soon as the young Euler started his schooling it was obvious that there was something special about him. He had an amazing memory that made learning languages trivial and an equally powerful ability to work complex calculations in his head with blinding speed. But these skills alone did not take him away from his apparent destiny in the pulpit. It was his experience when he began studying as a fourteen-year-old at Basel, the local university, in 1721.

Basel was, frankly, a backwater. Not the sort of place that

you'd expect to find great teaching staff. It turned out average students who went on to do uninspiring things. But one member of the faculty stood out head and shoulders above the rest. This was Johann Bernoulli, a top-class mind who would have been first to acknowledge himself as the greatest living mathematician. Euler was taking the opportunity to cover a broad range of topics including maths before entering the school of divinity to prepare for his calling to the ministry. Bernoulli, an old friend of Euler's father Paul (they'd shared accommodation when they were students together) saw merit in the youngster. Being Bernoulli, this was grudging. Euler observed:

> He was very busy and so refused flatly to give me private lessons; but he gave me much more valuable advice to start reading more difficult mathematical books on my own and to study them as diligently as I could. If I came across some obstacle or difficulty, I was given permission to visit him freely every Saturday afternoon and he kindly explained to me everything I could not understand.

Euler finally got to the stage where he should specialize and entered the school of divinity, but by now the appeal of mathematics was too strong. As he said:

> Not much progress was made [in *theology*], as I turned most of my time to mathematical studies, and by my happy fortune the Saturday visits to Johann Bernoulli continued.

In 1727, at the age of 20, Euler took a post at the brand new eighteenth century equivalent of MIT, the Academy of Sciences in St Petersburg. The idea of moving to Russia came from Bernoulli's son Daniel, who was already teaching maths there. Euler would be filling the position left by yet another Bernoulli, Daniel's brother Nicolaus, who died tragically young.

There was a snag, though. Euler certainly wanted to get away from the claustrophobic atmosphere of Basel. His conversations with Bernoulli had strayed from the purely academic – he had glimpsed the glamour of the wider world, a glamour that the city of St Petersburg had in abundance. But there were no jobs at the Academy in pure mathematics. All that was available was Nicolaus's post, which combined maths with physiology. Although he knew practically nothing about medicine, Euler managed to get the job thanks to the Bernoulli family's recommendation.

Once he had accepted, Euler's heart sank. It had been so thrilling to be offered a job, *any* job, at St Petersburg that he had given no thought to his ability to deliver, to teach a subject that he knew so little about. He put off his start date to the following year and began to read furiously, systematically acquiring as much knowledge of medicine as he could. It may have been that his physiological ideas owed more to geometry than was strictly healthy for any patients he might have to apply a knife to, but he had got the basics under his belt by the time he moved to Russia.

By now, Euler was reasonably certain he could stay one step ahead of his students. But the delay was not solely to enable him to read up on the subject. The chair of physics at Basel had become vacant and Euler was in the running. However, he didn't get the post (this isn't surprising considering his age) and he headed off for to the East. It wasn't an easy journey. It took him over a month, travelling down the Rhine by boat, across Germany with a post wagon and then again by boat to St Petersburg, but to the young Euler it was no more an inconvenience than backpacking around the world is to a modern student – it was all enjoyable experience.

Luckily for him, and for the students and patients he might have had to deal with, on his arrival at St Petersburg he discovered that a vacancy had come up working in physics, and recognizing his more appropriate experience, he had been assigned to that post without ever having to look at a body. Euler loved St Petersburg from the moment of his arrival. The cosmopolitan city was much more lively than Basel, and he found a good friend in Daniel Bernoulli. Just as their fathers had, they shared rooms together and enjoyed many common interests. They were young men with excellent positions and a burning desire to explore the world of mathematics. It was with some sadness, then, that Euler saw Daniel leave six years later – but any discomfort was tempered by taking over his friend's chair as professor of mathematics.

Soon after, Euler married and found that family life suited him to the ground. He commented later that he made some of his greatest mathematical breakthroughs while holding one of his 13 children in his arms (only five survived infancy). Although it was maths at a pure, theoretical level that really excited Euler, he was not above looking at practical applications too – in fact he had a very healthy interest in such a wide range of subjects that he enjoyed almost all the tasks he was given by the Russian government. He quickly became known as *the* man to call in – in the present day he would have spent most of his time on commissions of enquiry. His brief covered advising the navy, preparing maps for the government and acting as a consultant on fire engine design. He did, however, turn his nose up at one request – to cast a horoscope for the Czar. Euler was worldly enough not to irritate the ruler by bluntly turning down the request; he just made sure it was dealt with by somebody else.

Not long after his marriage, Euler suffered from a severe fever, and practically lost the sight in one eye from a resultant infection (though he blamed his loss of sight on too much time spent studying maps). Even so, once he got into his stride, Euler's output was prodigious, often contributing half the content of the Academy's journal. But his working life at the Academy was increasingly unpleasant. The head of the establishment was a miserable man by the name of Johann Schumacher. He was no academic himself, but a bureaucrat whose actions seemed designed to squash anyone showing signs of talent. At the same time, Russia's politics in the wake of the Empress Catherine's death were making life difficult for foreigners. In a letter, Euler commented that Russia had become 'a country where everyone who speaks out is hanged'. When an offer came from the King of Prussia to join his Berlin Academy, Euler didn't think too long. It was time to move again.

Letters to a German princess

Euler was to remain at the Berlin Academy for 25 years, dabbling in administration alongside his mathematical work. But despite his long stay, he faced a constant and growing opposition from an unexpected source – the king himself. When first appointed, Euler had been delighted to be singled out personally by King Frederick. He wrote to a friend: 'The king calls me his professor, and I think I am the happiest man in the world'. But Frederick was an intellectual groupie, with a love of the arts and particularly anything French, which soon became the official language of the Academy. Frederick's intent in recruiting Euler seemed to be to add to the sophistication of his court. But Euler was not particularly witty, nor had he the social graces that the king expected.

To make matters worse, Frederick's other star catch was the French author Voltaire. He was everything that Euler was not, and regularly made fun of the mathematician, spurring on Frederick's disdain for the man that he had previously courted so assiduously. Towards the end of Euler's stay in Berlin, the king's attitude was becoming intolerable. But Euler had never lost contact with St Petersburg, and with the accession of Catherine the Great, life had become much more stable in his old home. He returned triumphantly to Russia to remain until his death in 1783, working right up to the last despite going totally blind in 1771. It was typical of Euler that on the morning of his last day, he still managed to give a maths lesson to one of his grandchildren, made a series of calculations on the motion of balloons and held a spirited discussion on the implications of the newly discovered planet Uranus.

Euler's time in Berlin was anything but wasted. He produced two of his greatest mathematical proofs and a collection of letters that became his best selling work. His correspondence began after being approached by an envoy from an unlikely source. The Princess d'Anhalt Dessau, one of the Prussian king's nieces, wanted to hear about the latest developments in science. This would be incredible now – imagine a present day European princess (or perhaps a better parallel would be a female movie star) asking Lucasian Professor of Mathematics

Stephen Hawking to give her a taster of what's what in popular science. Back then it was little short of revolutionary, when the most a woman in her position would be expected to do was dabble in music and needlework. In fact, it's amazing that the whole business wasn't stopped as being dangerous to her health.

Luckily, Euler took a positive view of women's education, a sentiment echoed by Henry Hunter, who made an English translation of the letters in 1795. Hunter comments that they are translated for:

> the improvement of the female mind; an object of what importance to the world! I rejoice to think I have lived see female education conducted on a more liberal and enlarged plan. I am old enough to remember the time when well-born young women, even of the north, could spell their own language but indifferently, and some, hardly read it with common decency; when the young lady's hand-writing presented a medley of outlandish characters [...] They are now treated as rational beings, and society is already the better for it.

To add to the cosmopolitan flavour, the letters were written in the king's preferred language, French, and published as *Lettres à une princesse d'Allemagne* after Euler had returned to Russia. They became a best seller in English translation. After all, the princess was not alone in wanting to know more of the scientific revolution. Euler started gently, beginning with the science that underlay the princess's favourite subject, music. But it was when he got into his swing on the subject of light that he added his own contribution to that of Huygens.

Euler made a direct comparison of light to sound, which by then was well understood:

> The propagation of light in the ether is produced in a manner similar to that of sound in the air; and, just as the vibration occasioned in the particles of air constitutes sound, in like manner, the vibration of the particles of the ether constitutes light or luminous rays; so that light is nothing else but an agitation, or concussion, of the particles of ether, which is every where to be found, on account of its extreme subtlety, in virtue of which it penetrates all bodies.

Convinced that the Sun and other bright bodies were vibrating like a bell (this way they could give off light without disappearing – otherwise, Euler comments, the Sun must be speedily exhausted), Euler went further astray when explaining how light allows us to see things. Although he was happy that reflection could happen in a mirror – light bouncing off the surface just as a ball does – he thought a different process had to be occurring when an opaque object like a building was illuminated; otherwise, he argued, everything you saw would be covered with mirror-like reflections. Instead, he thought, the vibration of light started a sympathetic vibration in the object it lit, the way a piano string will start moving of its own accord when a note is played on a nearby instrument. What we then see, Euler thought, was the light emitted by the object's own vibration, not the original light from the Sun.

Euler may have got it wrong when it came to the way we see objects, but he also tightened up Huygens' maths, adding practical detail to describe the way in which waves move through a body like the ether. His contradictions of Newton's theories were always carefully argued. But Euler and the other supporters of the wave theory were not the only ones to dispute Newton's supremacy. A more unlikely challenger was the great German writer Johann Wolfgang von Goethe.

Perception and reality

Though Goethe is widely recognized as a literary giant, his interests spread far beyond the written word. From his time at the University of Strasbourg in the 1770s, Goethe did not limit himself to letters and philosophy. Despite his prodigious output of plays, novels and poetry, his degree was in law, and he found time to explore music, art and science.

By his fortieth birthday, Goethe's literary position should have been unassailable. He had just spent three years leave of absence in Italy and returned to Weimar, the capital of the duchy of Saxe-Weimar, expecting to slip comfortably back into the aristocratic circles in which he had moved before his trip abroad. But he found unexpected resistance. His literary approach, which had moved from the dramatic emotional torment of *Sturm und Drang* to a more classical, contemplative style while in Rome, was ironically considered too modern. At the same time he risked exclusion by his peers for taking up with a 23-year-old girl, Christiane Vulpius. At the time a flower factory worker, Christiane was soon to bear his first (and only surviving) child. Seven years later, in 1806, they were married.

Goethe's response to this unhappy return to Weimar was a form of midlife crisis, a temporary shift of allegiance. Though he would return to writing refreshed, and later produced his masterwork, the dramatic poem *Faust*, for a while it was science that bore the full weight of his attention. Just as his writing had concentrated on the relationship of individuals to nature and society, his subjective brand of science brought the human component to the fore. Much of this study was dedicated to biology, but alongside it was a desire to uncover the secrets of light, and particularly colour.

It nearly wasn't so. Goethe had, in a fit of enthusiasm, borrowed a box of optical equipment from an acquaintance, Hofrat Buttner of the German town of Jena, which would become famous as the home of the Zeiss optical works. The box arrived at an inconvenient moment; it was pushed out of the way and quickly forgotten. As time went by, Buttner became impatient with Goethe. He sent a messenger to collect his equipment. Goethe, who had not got round to even opening the box, told the messenger to wait for a moment.

He lifted the lid on the battered wooden case and rummaged through the contents. Near the top was a large glass prism, wrapped in a soft sheet of velvet. Goethe took the prism out, enjoying the feel of the heavy glass in his hands. He couldn't resist putting the prism into the rays of the Sun from his window. No colours emerged. He twisted the prism about. Still no colours. In bizarre triumph, he cried out 'Newton was wrong!' and called in the messenger.

This little incident sparked off an interest in light and colour that would last for the remaining 40 years of his long life. His efforts always seemed allied to an urge to disprove Newton's findings. Quite what Goethe had against Newton isn't clear (perhaps that he was more famous than Goethe himself), but there was certainly a determination to undermine the great man's work.

Goethe wasn't up to the technical rigours of showing that

Newton's corpuscles, his tiny particles of light, didn't exist. But Goethe's brief error (he soon managed to produce a spectrum when he bought his own prism) sent him off in a direction that should have been more compatible with his talents. Like Leonardo da Vinci before him, Goethe displayed different colours alongside each other. It soon became obvious that the colours changed depending on context. A bright red, for instance, looked different if put alongside dark blue or pale pink.

Technical theory might not have been Goethe's strongpoint, but he had plenty of persistence. He experimented with a vast range of colour combinations. The more he researched, the more he was sure that he had found the weak point in Newton's analysis of colour. Newton had said that any particular colour was an absolute, fixed property of a certain light, the light that was bent by a particular amount by a prism. Yet Goethe could take a slice from the output of a prism and put it alongside various other colours. When he did so the colour of the prism's light changed.

In the subjective, almost poetic terminology that he employed, Goethe tried to show that colour was not an absolute, but depended entirely on the nature of the stimulus. A particular light, he argued, didn't have a fixed colour; the colour depended on how and where the light was seen.

In keeping with his classical human-centred view, Goethe had made a classic mistake. He was confusing what his senses told him with reality. In effect, both Goethe and Newton were right, but Newton was describing what *light* was like, while Goethe was describing what *the human perception of light* was like. Had this been his intention, Goethe's work would have stood up as a useful contribution to the sum of scientific knowledge. Unfortunately, he had, like Newton, been attempting to describe light itself. The result was inevitably a muddle.

Colour is an absolute property of light, but the eye's mechanism for determining which colour it sees does not solely rely on the appearance of the single colour. Eyes are not designed for picking out individual colours, but for using colour to distinguish shapes and objects. For this purpose, relative colour is more important than absolute. The eye registers a colour's relationship to its surroundings, not a pure value. This practical compromise unfortunately misled even as great a thinker as Goethe.

Despite Goethe's attempts to counter them, Newton's theories continued to dominate scientific thinking. Huygens' and Euler's championing of waves went practically unheard until a triumvirate of British scientists brought Newton crashing down from his apparently unassailable position.

Chapter Six

Light's anatomy

*We all know what light is; but it is
not easy to tell what it is.*
SAMUEL JOHNSON – BOSWELL'S LIFE

The battle between Newton's and Huygens' competing theories was to be settled by an unlikely figure. Thomas Young was a doctor who never worked full time in physics. He was the epitome of the enthusiastic amateur, trying his hand at everything that interested him. Young worked on botany and physiology, brought the concept of elasticity to engineering and produced mortality tables to help insurance companies set their premiums. He made the first translation ever of ancient Egyptian hieroglyphics, contributed ideas to philosophy and still managed to keep his medical practice in London going.

A true polymath

This breadth of interest started early. Young, who was born at Milverton in Somerset in 1773, taught himself to read at the age of two. His parents only discovered this when he came to ask for help with some of the longer words in the Bible. The Young house, which still stands in North Street, Milverton, became increasingly crowded as nine siblings joined young Thomas. He spent much of his early years at his maternal grandfather's in Minehead, where a large library helped him expand his horizons.

While at boarding school, Young picked up new languages so easily that he was called on to demonstrate his skills as an amusing novelty for visitors. By 13 he fluently read Greek, Latin, Hebrew, Italian and French. But after a period of private education it was the great teaching hospital St Bartholemew's in London and then Cambridge University that developed his medical and scientific interest. With the final, more worldly contribution of £10,000 and a house in London from a legacy, Young was set up for a life of amateur discovery. In the spring of 1799 he opened an office at 48 Welbeck Street. Within two years he would be at the centre of the greatest scientific controversy of his day.

Young was among the last of those who could be an amateur and still make a major contribution to science. He may have seemed a dilettante to some of his contemporaries, but whatever he applied himself to he excelled in. As his epitaph in Westminster Abbey says, he was 'a man alike eminent in every department of human learning'. Perhaps Young's greatest gift was making intuitive leaps. By 1800 he had followed Huygens in accepting that light was a wave. What he lacked was an experiment to prove his theory. The breakthrough was to come entirely by accident.

Young was studying the effect of temperature on the formation of dewdrops, shining a candle light through a fine mist of droplets. The images of these droplets, projected on a white screen, formed coloured rings around a white centre. Young suspected the rings were caused by the waves of light interacting with each other. Inspired by this discovery, he spent many hours in a darkened room, channelling beams of light around a table. In 1801, he gave a lecture to the Royal Society called

The Theory of Light and Colours. Young was determined to show that, for once, the great Newton had got it wrong. His secret weapon in this attempt was an odd effect he called interference.

Waves interfering

Young had shone a tight beam of light onto two closely spaced slits in a piece of card and then allowed the resultant twin beams to fall on a piece of paper behind the card. It might seem reasonable that the result would be a bright portion in the middle where the light from the two slits overlaps, a pair of dimmer sections either side lit by a single slit alone, and then darkness at the edges. Instead, Young saw a row of narrow alternating bright and dark bands.

There was no obvious reason why this should happen if Newton's picture of streams of tiny particles were true, but Huygens' waves were much more promising. When two waves meet, they don't ignore each other. If both waves ripple upwards at the same time you end up with a wave that is twice as big. If one is rippling up at the same time as the other is rippling down, the two cancel each other out, leaving no motion at all. You can see this happening if you drop two stones near each other in a pool of water. Where the waves overlap there will be points where the water hardly moves and others where the motion is particularly strong. Young argued that exactly the same thing was happening to light after it passed through the two slits. The waves of light were interfering with each other. As they moved out from the slits, at some points they were both rippling up at the same time, producing the bright bands. At others they were cancelling each other out, giving the dark bands (Figure 6.1).

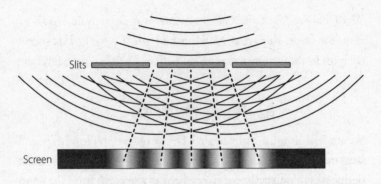

Slits

Screen

Dotted lines show where waves reinforce,
producing bright patches on screen

Figure 6.1 Light waves interfering with each other after passing
through Young's slits

For good measure, Young threw into his lecture a link
between waves and the different colours of light. To understand
this takes a few moments' thought about what a wave is. The
up and down motion as the wave moves along can come slowly
or quickly. If you imagine a wave – a ripple in a skipping rope,
for example – moving at a fixed speed, the more ups and downs
there are, the closer together they will be. Think, for instance,
of a sewing machine needle diving up and down through a
piece of cloth as the cloth moves steadily along. The faster the
needle makes up and down movements, the closer together
the stitches are. The distance in which a wave passes through
a whole up and down movement, returning to the same point
in the wave is called its wavelength (Figure 6.2). The number
of these wave movements that take place in a second is its frequency. Young found that the patterns produced by his slits
varied as he changed the colour of the light. The alteration was
exactly what would be expected if the wavelength was changing
as he changed the colour, making the interference of the waves

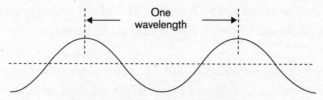

Figure 6.2 The wavelength of light

happen closer together or farther apart. The colour of light, he deduced, was directly connected to its wavelength.

As Euler had before him, Young initially thought that the waves of light, like sound, consisted of a compression in the same direction as the wave moved. The result would be like the squeezing and releasing of the bellows of an accordion or a sudden push passing down a 'slinky' spring. This seemed natural because it was thought that light travelled through the ether, an invisible, intangible fluid filling all space, and you can't send up and down or side-to-side waves through the middle of a fluid. But hearing of Fresnel's work on polarization (see page 124), Young made a bold suggestion that was to open him to ridicule. The obvious explanation for the way polarized light behaved was that light's waves did in fact move up and down or side to side like the ripple in a rope. Young couldn't think of any mechanism that would allow this to happen, but it was the only suggestion that seemed to fit the facts.

Although Young's work was a huge advance on anything that had previously been known about light, and his arguments were both simple and powerful, his view was not widely accepted for another 40 years. In England particularly he was ridiculed for opposing the indomitable might of Newton's legacy. He received vitriolic criticism from the establishment, especially Henry Brougham, at the time a young lawyer and writer and later Lord Chancellor. Brougham was a founder of the influential *Edinburgh Review*, in which he wrote:

> We may now dismiss for the present, the feeble lucu-brations of this author, in which we have searched without success for some traces of learning, acuteness or ingenuity that might compensate his evident deficiency in the powers of solid thinking, calm and patient investigation, and successful development of the laws of nature by steady and modest observation of her operations. Has the Royal Society so degraded its publications into bulletins of fashionable theories for the ladies of the Royal Institution? Let the Professor continue to amuse his audience with an endless variety of such harmless trifles, but in the name of Science, let them not find admittance into the venerable repository which contains the names of Newton, Boyle, Cavendish...

On the continent, though, Newton was held in less reverence, and Young's wave ideas were to be bolstered by Augustin Jean Fresnel (totally unaware of Young's work) who showed how the fringes that Grimaldi had discovered around a shadow, now called a diffraction pattern, could be explained by exactly the same mechanism as Young's interference.

The road-builder's triumph

Fresnel was a very different character from Young. Although aristocratic and haughty in appearance, Fresnel did not have the cachet of some of his more cosmopolitan colleagues, and

certainly could not afford the playboy life of a scientific dilettante. He was a practical man, an officer of the Corps de Ponts
& Chaussées whose day-to-day work concerned the utilitarian
business of building roads and bridges. What's more, he had
shown himself to be anything but politically astute, opposing
Napoleon's return from Elba, which earned him a short spell
in prison. Just as Euler had been snubbed by the Prussian king,
the down-to-Earth Fresnel was dismissed by the Parisian scientific establishment.

Even when Fresnel presented his carefully devised equations, showing how waves of light could produce the fringes at
the edge of shadow, there were those who could not accept that
a mere road-builder could be capable of such original thinking. Fresnel ignored them and worked systematically on with
his investigations, unknowingly echoing Young in getting help
from the village blacksmith in constructing his apparatus. But
while experiments were at the heart of Young's work, Fresnel
was a more accomplished mathematician and could back up his
theories with a polished analysis.

One of Fresnel's critics was the mathematician
Simeon-Denis Poisson. Poisson might have moved more in
the right social circles than Fresnel, but he also knew his stuff,
making wide-ranging contributions to science and maths as
the first holder of the chair of mathematics at the Sorbonne.
Poisson found Fresnel's work ludicrous. To show how absurd
his fellow Frenchman's calculations were, Poisson predicted,
using Fresnel's equations, that there should be a small spot of
brightness directly behind a solid object in the path of a light
beam. This, said Poisson, was clearly ridiculous, the imagining of a feeble intellect. But Poisson's attempt to mock Fresnel
backfired. Another successful scientist, Dominique François

Jean Arago, put Poisson's ridicule to the test. He shone a light at a small target and found a bright spot behind it, exactly where Fresnel's equations predicted it would be. The road builder was vindicated.

By now Newton's theory was like a punch-drunk boxer: still standing, not realizing that he has already been knocked out. Young and Fresnel had made the particle theory untenable; from now on the only acceptable view of light would be as a wave. But the time was approaching to go far beyond the simple argument of waves or particles. The next great challenge was to discover how light worked, what made it travel at such an immense speed – what it truly was. It took a remarkable man and his terrified friend to give the first, unexpected glimpse of light's innermost secret.

The accidental lecturer

It is possible to be unusually precise about the timing of this breakthrough. Just before nine o'clock on the evening of Friday 10 April 1846, two men waited in the wide lobby behind the stage doors of the Royal Institution in London. They were Charles Wheatstone and Michael Faraday, both respected physicists. Wheatstone was due to give a lecture on his electromagnetic chronoscope, a novel, electrically controlled clock. But the pressure of appearing in front of these Friday evening audiences fell heavily on a nervous speaker like Wheatstone. Faraday, who had set up the lectures, had begun a tradition kept up to this day that the speaker should rush straight onto the stage and begin his topic without pause or pleasantries.

Nerves finally got the better of Wheatstone. Dropping his notes, he rushed out of the building, leaving Faraday in a difficult position. The audience was already assembled. It would be embarrassing for the Institution to cancel at so late a stage. Faraday collected together his friend's notes and skimmed through them. He knew that he could give the lecture – no doubt better than his panicky friend – but he could not make the material last for a whole evening. He would have to add something of his own. Without time for preparation, Michael Faraday was about to give the most inspired lecture of his career: a first insight into the inseparable nature of light, electricity and magnetism.

Forty-one years earlier, Faraday's family had been driven from Westmoreland to London in a desperate search for work. Michael had little hope of being anything more than a blacksmith, like his father. But at the age of 14 he was apprenticed to bookbinder George Riebau, a refugee from the French revolution. It was Faraday's turning-point. Riebau encouraged his apprentices to learn more than just bookbinding; Faraday spent all his spare time in the shop, soaking up the contents of the books around him. These heavy volumes, soon his closest friends, and the lectures of a self-improvement group, the City Philosophical Society, gave Faraday a single-minded purpose. He intended to break into the closed world of science.

In a twist worthy of a Hollywood movie, Faraday would soon get within touching distance of his ambition, only to have it snatched away. George Riebau showed Faraday's carefully bound lecture notes to a client, Mr Dance. Dance and his father were so impressed that they sent Faraday tickets to attend Humphry Davy's lectures at the Royal Institution. The unexpected gift thrilled Faraday; Davy was a Victorian scientific superstar. But this opportunity was only the start. Soon after, an experiment went wrong for Davy. The equipment exploded in his face, blinding him. Mr Dance dropped a hint that Faraday would make the ideal secretary and Davy took him on.

Davy's protégé

Working with Davy was a dream opportunity, but it did not last long. Unfortunately for Faraday (if not for Davy) the blindness was only temporary. As soon as Davy could manage alone, Faraday was sent back to the bookbinders. To get so close and then be rejected could have been devastating, but Faraday's single-minded drive and enthusiasm saved him. He kept up a steady barrage of applications for jobs in scientific establishments. Even so, in the end it was another man's drunkenness that succeeded where Faraday had failed. The lab assistant at the Royal Institution, William Payne, was sacked for brawling. Faraday filled the empty post with Davy's blessing. Davy's assessment of Faraday was recorded in the minutes of the Institution:

> Sir Humphry Davy has the honour to inform the managers that he has found a person who is desirous to occupy the situation in the Institution lately filled by William Payne. His name is Michael Faraday. He is a youth of twenty-two years of age. As far as Sir H. Davy has been able to observe or ascertain, he appears well fitted for the situation. His habits seem good; his disposition active and cheerful, and his manner intelligent. He is willing to engage himself on the same terms as given to Mr. Payne at the time of quitting the Institution.

By 1821 the young scientist's life was steady and unremarkable. Faraday had been promoted, and with the extra money could afford to marry Sarah Barnard, another member of his strict religious community. With her he moved into the suite of rooms in the Institution that had previously been home to Young and then Humphry Davy. He was anything but controversial. Yet disaster loomed. Asked to write an article on electromagnetism, the interplay of electricity and magnetism, Faraday was delighted to get his hands dirty. He repeated the experiments he had read about, not prepared to accept the results without seeing them for himself. As he passed electricity down a wire running alongside a fixed magnet, he saw something that first puzzled him, then filled him with excitement. The wire moved of its own accord, circling round the magnet. No one else had mentioned this; it was Faraday's discovery. With unusual haste for such a careful man he rushed to publish his findings and sat back to await the applause. It didn't come. Instead, he was accused of plagiarism.

The attack came from William Wollaston, one of the scientists whose work Faraday had reviewed. Wollaston was originally a doctor, but partial blindness drove him to give up medicine. He had dreamed up an unlikely idea that electricity ran along wires in a corkscrew spiral. Wollaston persuaded his friend Sir Humphry Davy to help him search for signs of this movement. Their attempts failed, and Wollaston's theory bore no resemblance to Faraday's experiment, but it was enough for Wollaston that there was electricity and rotating movement involved. The finger was pointed at Faraday. Shocked by the suggestion that he had broken his cast-iron moral code, Faraday turned to Humphry Davy for support. Instead he was abandoned.

Davy chose to side with his friend Wollaston. For all his apparent acceptance of Faraday, the class divide was too great. Davy was a society hero, mixing with royalty. Faraday, as far as he was concerned, would always be an upstart. It was telling that when early in his career Faraday accompanied Davy and his new wife on a tour of European scientific establishments he was expected to act as valet as well as scientific assistant. Wollaston, on the other hand, was a professional man, 'one of us' in Davy's mind. Davy had given Faraday every opportunity to rise from the gutter – now this had happened. The rift was permanent. The two men never exchanged a friendly word again.

Before long it became obvious to everyone that Faraday's discovery was original. Not only that, but immensely useful. The steady motion of the wire around the magnet was the basis of every electric motor. Faraday's name was made. When two years later he was elected to a fellowship of the prestigious Royal Society, only one person voted against him. Sir Humphry Davy.

Lines of force

What no one realized in 1821, least of all Faraday, was that his fascination with electricity and magnetism would lead to a breakthrough in our understanding of light. But it would be ten years before Faraday returned to electricity and magnetism. The pain of the accusations and Davy's betrayal bit deep. He turned his attention to chemistry and took on the administrative job of Director of the Laboratory, establishing the nine p.m. Friday lectures and a series of Christmas events for children. Both still operate – Faraday's Royal Institution Christmas lectures are a familiar sight on British television.

Successful though his chemical ventures were, Faraday could not resist the challenge of electromagnetism forever. By 1831 there had been hints that electricity flowing through a wire could generate a current in another, unconnected wire, somehow communicating across space. This near-magical proposition was too much to resist for a man with Faraday's curiosity. He rigged up a pair of wire coils, wrapping each long piece of wire around the straight sides of an elongated hoop of iron. He expected to see a steady flow of electricity in the second coil, somehow leaking through the metal hoop, when he powered up the first. Instead there was only a brief surge of power in the second coil when the first coil was turned on or off.

A lesser man would have blamed his equipment or dismissed the evidence, but Faraday worried away at the problem like a dog tugging at an old sock. It seemed unreasonable that just switching on and off the electricity in the first coil of wire influenced the second at a distance. But it was already well known that a coil of electric wire could produce magnetism. And magnets certainly did work at a distance – a compass proved that. So what if the first wire was acting as a magnet?

This was the inspiration Faraday needed. If it was the changing level of magnetism that generated a new current, rather than electricity leaking across, it made sense that there was only a short burst of current in the second wire. Before long he was producing electricity by moving a normal magnet through a coil – he added the generator to his list of inventions. When British Prime Minister Robert Peel asked Faraday what use his new discovery was, he is alleged to have replied: 'I know not, but I wager one day your government will tax it'.

The crossover between electricity and magnetism left scientists struggling for a language to describe what they saw. A popular party trick at the time was to sprinkle iron filings on a sheet of paper held above a magnet. The tiny shards of metal cluster together in curved lines that seem to map out the magnet's invisible power. Sitting in the dim evening light of his laboratory, Faraday imagined these magnetic lines glowing in the air. When he moved a wire near the magnet, the wire was hitting the glowing lines, as if he was running along like a child beside iron railings, slapping them with his hand. Each slap generated power as the wire hit and cut through those imagined brightly glowing lines. Like the shock of the slap running through his arm, an electrical current would run through the wire. He called these railing lines that surrounded the magnet 'lines of force'.

If this was the case, Faraday, mused, what had happened when he switched on his electrical coil? Then there was a brief surge of electricity in the second wire. It was if the lines of force – the bars of the railings – were all bunched up in the coil. When he switched it on, turning it into a magnet, the lines moved out into place. As they moved, they cut themselves on the wire, as if the hand was held still and the railings were sliding past it, doing the slapping themselves. This led Faraday into deeper considerations. The lines of force didn't jump instantly into place as soon as he switched on the coil magnet; they took time to move out into position, otherwise the wire wouldn't be cutting through the lines. Something was travelling through the air, some invisible magnetic phenomenon.

This was a brilliant observation, but Faraday was wary about telling anyone. He remembered the pain when Davy had abandoned him, the feeling that his honour and integrity were under question. Rather than publish his results, he hid his ideas away in a sealed envelope, dated 12 March 1832, to be opened after his death. Concealed in the darkness of the safe were the first hints of Faraday's remarkable speculation about light. The lines of force, the iron railings of magnetism, moved out from the electromagnet when he switched it on. What exactly was moving? He wrote:

> I am inclined to compare the diffusion of magnetic forces from a magnetic pole, to the vibrations upon the surface of disturbed water, or those of air in the phenomena of sound: i.e. I am inclined to think the vibratory theory will apply to these phenomena, as it does to sound, and most probably to light.

Going public

Faraday's inspired linkage of magnetic vibrations – waves – and light lay buried in the safe inside its sealed envelope until 10 April 1846, when the clock finally ticked to the hour of nine and Faraday took Wheatstone's place on the stage. He had given a few hints the year before, pointing out that:

> I have long held an opinion […] that the various forms under which the forces of matter are made manifest have one common origin […] This strong persuasion extended to the powers of light.

Faraday was pointing to some linkage between the forces of electricity and magnetism and light. But it was not until that remarkable evening that he brought his speculation into the open. There is some suspicion now that the story of Wheatstone's panic is a myth. The Royal Institution's records show that Faraday substituted for another scientist, James Napier, who gave a week's notice of his absence. It is certainly true, though, that Faraday spoke about Wheatstone's delightfully named but wholly forgettable electro-magnetic chronoscope. When his colleague's notes ran out, all too soon, Faraday took a deep breath.

Perhaps it was the impromptu nature of the occasion that lowered his guard. Perhaps, unlike 15 years earlier, he now felt that he had status enough to take the risk. Unprepared and without a safety net, Faraday began to speak his mind.

He described light as a vibration, rippling through the iron railings of the magnetic force lines. To fully understand just what an amazing insight this was, you have to put yourself in the audience, in that lecture theatre in 1846. It was a time before electric lighting, when oil lamps and candles and a few gas lights were the only sources of night-time illumination. To Faraday's audience, electricity and magnetism were still very new, the power behind machinery like the chronoscope. Faraday's leap of genius, connecting the ethereal phenomenon of light with magnets and electrical coils, was inspired.

He let his imagination run free. He later said that he 'threw out as matter for speculation, the vague impressions of my mind'. The outcome was stunning.

The views which I am so bold as to put forth consider, therefore, radiation as a high species of vibration in the lines of force which are known to connect particles, and also masses of matter, together. It endeavours to dismiss the aether, but not the vibrations.

In one amazing step Faraday pointed to the nature of light and eliminated the need for the 'ether' – the substance that was assumed to fill empty space to allow light to travel through it, just as water or air carried sound waves. It would be more than 50 years before the ether argument was settled definitively, but Faraday had advanced the first mechanism for light that didn't need it. Light, he thought, *was* a wave, but a very different wave from sound.

> [...] the vibrations [*of sound*] are direct, or to and from the centre of action, whereas the former [*light*] are lateral.

Building on Thomas Young's largely rejected ideas of a generation before, Faraday suggested that where sound moves along by squashing and releasing air, like motion along the bellows of a concertina, light moves in a side-to-side ripple. He even went so far as to suggest that gravity acts in a similar way, an inspiration that it took Einstein to develop further.

Faraday's vision was dazzling, but he retained a humble view of his own importance. He refused a knighthood (unlike his friend Wheatstone) believing that any credit was owed to God, not him. Yet his sheer enthusiasm for science and his ability to see beyond the obvious made the modern approach to light possible. It only took the genius of Scotsman James Clerk Maxwell to build on Faraday's foundations and open up light's true nature. But before this revelation, an accurate measurement would be made of light's speed. Two Frenchman, Armand Fizeau and Jean Bernard Léon Foucault were determined to catch up with the fastest thing in the Universe.

Timing light

Most of the ancients held that light travelled instantaneously from place to place. Although Empedocles, with his popular idea of light streaming from the eye, did think it moved at a measurable speed, he was overruled by Aristotle, who described light as a state of the medium it travelled through. The medium, he thought, switched into 'light mode' all at once, just as a pool of water can ice over in an instant. But he gave Empedocles some credit for his reasoning before dismissing it:

> Empedocles says that light from the Sun gets to the intervening space first before coming to the eye, or reaching the Earth. This might plausibly seem the case. For whatever is moved through space, is moved from one place to another. So there must be a corresponding time period in which it is moved from the one place to the other. But any given time is divisible into parts, so we should assume a time when the Sun's ray was not yet seen, but was still travelling in the middle space.

As long ago as 1676 Danish astronomer Ole Rømer showed that Empedocles was on the right track. He wasn't the first to doubt Aristotle, though. Both Alhazen and Roger Bacon (see Chapter 3) were sure that light must take some time to travel. Before Rømer was born, Galileo had tried to investigate light's speed with an experiment that was a triumph of hope over practicality. Realizing the difficulties of accurately comparing measurements on two clocks separated by a distance, Galileo devised an experiment in which light returned to its original source, needing only one clock to time it.

The night in the countryside around Padua was stygian as Galileo and his assistant set out to make their measurement. It's hard to appreciate just how absolute the darkness was when looking back from the present. Now the sky glow of artificial light reaches most of our world, but this was the pure black night of the Italian countryside in the seventeenth century. The assistant rode off a measured distance and stationed himself ready for Galileo's signal. Taking account of the clock, the great man unmasked his lantern, adding a yellow-white star to the view of his assistant. Immediately, the assistant uncovered his own lantern, and light was sent on the return journey, ready for Galileo to spot it and mark the time. The result was a disaster. There was no consistency in timing. Galileo returned home a failure. He commented that he had found it impossible:

> to ascertain with certainty whether the appearance of the opposite light was instantaneous or not; but if not instantaneous, it is extraordinarily rapid.

For once, the man whose faith in the invincibility of science allowed him to take on the hierarchy of the church was frustrated. Even if his timepiece had been accurate enough to measure the time light takes to travel that sort of distance – perhaps a 100,000th of a second – the delays introduced by human response times at both ends of the experiment far outweighed anything else. But at least Galileo tried. There weren't many natural philosophers of the time who would even dare to imagine that light had a measurable speed. The philosopher Descartes was one of the strongest supporters of this theory, writing:

[Light] reaches our eyes from the luminous object in an instant; and I would even add for me that this is so certain, that if it could be proved false, I should be ready to confess that I know absolutely nothing about philosophy.

Sixteen years after Descartes' death, in 1676, an obscure Danish astronomer Ole Rømer proved him wrong.

A long enough measure

Ironically it was a success of Galileo's that made Rømer's measurement possible, a measurement that Rømer never intended to make. He was working at the University of Paris, timing the four largest moons of Jupiter as they moved in and out of the giant planet's shadow. His hope was to provide navigators with a natural clock to pinpoint time while they were at sea.

Calculating the position of a vessel depended on accurate timekeeping, impossible onboard ship with the crude mechanical clocks of the day. Year by year, ships had been dashed on the rocks and destroyed, led astray by imprecise measurements. Ever since Galileo's discovery of Jupiter's moons in 1610, astronomers had been trying to map the regular movements of these distant points of light to use them as a precision timepiece. But Rømer discovered that the moons did not behave as he expected. The timings did not stay constant, growing later by the day. Rømer was curious. Why did the moons seem to be slowing down?

It was only the fact that his measurements were taken over a long period that gave Rømer the chance to realize what was happening. Day by day, the timings shifted, until one day the change was reversed. Now the timings became earlier every day. Rømer noticed that the change coincided with the time when Earth was at its most distant point .from Jupiter.

It was too big a coincidence to be unconnected. Rømer knew that as the Earth and Jupiter described sweeping arcs through the solar system, the distance between the planets grew bigger, reached a maximum, then grew smaller again. While the distance to Jupiter was increasing, the light took longer to reach him each day because it had further to travel. This extra time taken for the light to arrive made the appearances and disappearances of the moons seem later than they really were. Rømer only had to compare the change in timings with the change in distance to work out light's speed.

With the help of fellow astronomer Cassini's measurements of Jupiter's distance from the Sun, Rømer came up with a figure for the speed of light. At 220,000 kilometres a second he was almost a third too low, but the inevitable progress towards an accurate figure had begun. Over the next 300 years, closer and closer measurements of light's speed would be made until 1983 when, bizarrely, it ceased to be possible to make a measurement. Light's speed would never need to be checked again.

While Rømer's measurements were improved on, there was something not quite satisfying about using the remote movement of heavenly bodies to pin down light's speed. It would only seem truly in humanity's grasp if the measurement could be brought down to Earth. The man to do it was French, Armand Hippolyte Louis Fizeau. His idea was in many ways similar to Galileo's – it involved measuring the time for a beam of light to make a round trip to a distant point and back again. But how was he to get over Galileo's problem of timing a journey that would only take a fraction of a second? It was 1849, a very different era from Galileo's. Rather than rely on human reactions, Fizeau, a product of the mechanical age, opted for a mechanical solution.

Fizeau arranged for a bright light to be shone towards a mirror around nine kilometres away. Shortly after the light left the lamp, its tightly focused beam played on a wheel with hundreds (720 to be precise) of tiny teeth cut in the edge. As this wheel was turned, a series of flashes were sent off down the 9 kilometre track to the mirror and back. On its return, the light once again passed through the toothed wheel. Now came the clever bit. If the wheel was turning at just the right speed, it would have rotated by the width of a tooth in the time the light was on its journey. The light would be blocked by the tooth, and nothing would be seen. Let the wheel go a little slower or a little faster and the light would get through the gap either side of the tooth. By running the wheel at nearly the right rate, then varying the pace a little, a good measure of light speed should be possible.

It was so much simpler than Galileo's proposition. However accurate a clock was, the problem was managing to say 'now!' when the light started and stopped. This way, all that was necessary was to distinguish light from darkness. With 720 teeth and nearly 18 kilometres of path, it wasn't necessary to spin the wheel too quickly, either. About ten turns per second was all it took – a perfectly practical speed (compare this with the 100 to 150 turns per second of a modern computer hard disk), and capable of accurate measurement with the technology of the day.

Even Fizeau's experiment didn't quite put everything under the experimenter's control. The light still had to pass down his nine kilometre racetrack. But the next year a colleague, Jean

Bernard Léon Foucault, took the final step. Best known for his remarkable pendulum that demonstrates the rotation of the Earth, Foucault took Fizeau's 18 kilometre round trip and condensed it to 20 metres. With a handful of mirrors he contained this trip into a wooden apparatus about three metres long. On such a journey, there was much less time to react. A wheel to cope with the timings would have to have an impractical number of teeth and be rotated impossibly fast. But Foucault knew how to get round this – it was all achieved with mirrors.

On its way into Foucault's elegant brass and wood device, the light was bounced off a mirror that was being spun around at high speed by a stream of compressed air playing on a turbine. It then rattled around its path and returned to the same mirror, which had, by then, turned a tiny fraction. The new beam went back along the length of the apparatus. Because of the shift in the mirror, the beam hit the far end at a slightly different point from the entry of the original ray of light. By using a microscope and a very fine graduated scale, this shift could be measured. All that was needed then was to know how fast the mirror was going round, which could be achieved using a similar technique to Fizeau's with a toothed wheel. The speed of light was pinned down to 298,000 kilometres per second.

Death of the mechanical era

From Foucault, Albert Michelson (see page 190) took up the challenge and using developments of Foucault's device had pushed the figure to 299,774 kps by 1931. But Michelson had reached the limits of what was possible with a purely mechanical approach. Something different had to be used. The solution to this problem depended on the peculiarities of waves. There are three measurements you can take of a wave. Given any two of them you can come up with the third. They are the speed, the wavelength (the distance it takes the wave to go through a whole ripple and get back to the same position in its up and down cycle) and the frequency (how often it ripples each second). Multiply the frequency and the wavelength together and you get the speed. Ever since the 1950s, this simple relationship has provided ways of measuring light's speed with greater and greater accuracy.

By using features of waves that are familiar to musicians – the way a box of a particular size will react to a particular frequency, just as an organ pipe does, the way any particular frequency is usually accompanied by higher multiples of that frequency called harmonics, and the way two very similar frequencies will interact with each other to produce a very slow wave called a beat – it was possible to pin down the wavelength and frequency of different types of light (and hence its speed) to remarkable accuracy. By the early 1980s, light speed was measured at between 299,792,457 and 299,792,459 metres per second. But this presented a new problem.

The second had been defined for some while by an atomic clock, measured to incredible accuracy by the unchanging oscillation of a caesium atom as it flicks between different levels of energy, but the metre was defined by the wavelength of the light emitted by a krypton atom, and this measurement wasn't as accurate as that of light's speed. So light speed, measured in metres per second, was known more accurately than exactly what a metre was. Something had to give. In 1983 it was decided to fix the speed of light once and for all as 299,792,458 metres per second.

At first glance this seems impossible. How can we

arbitrarily give an exact value to something that is a property of the Universe, not a manmade concept? Because the metre was redefined in terms of light speed. The metre is now 1/299,792,458th of the distance light travels in a second. As measurements get better and better, our idea of what a metre is will subtly change, but light speed is fixed forever. At least we hope so. In 2000 a Portuguese physicist based at Imperial College London published a paper suggesting that the speed of light, this fundamental constant value, might not always have been the same.

A true constant?

When cosmologists look back in time to the origins of the Universe there are some uncomfortable inconsistencies. Assuming that there was a big bang at a fixed point in the past, now thought to be about 13.7 billion years ago, the furthest light could have travelled across the Universe is the distance light can cover in that time at 300,000 kilometres per second. But there is some evidence that it got further. Parts of the Universe spread further from each other than that have a uniformity that suggests they were once in touch – but how was this possible when the fastest thing in existence couldn't cross the gap?

The most common explanation is inflation – that somehow in the earliest times, space itself expanded like a balloon, putting in the extra distance that made it impossible for light to cross in time. But Doctor João Magueijo has suggested a very different proposition. What if light moved much faster in those early days than it now does? This 'variable speed of light' theory is not generally accepted, but equally has not been proved wrong. It's just possible that this most fixed of values was not always so. Perhaps the sort of tunnelling we explore in Chapter 9 was happening then. Or perhaps it was time itself that did not flow as we now understand it, so near the beginning. Cosmologically fascinating, these speculations don't alter the facts about light as we know it today. Light's speed is not going to change.

In love with colour

If it were possible to stand back from history and take in the whole of the development of humanity's slow struggle to understand light, one man would stand out above all others. His is not a household name. He certainly hasn't had the fame that he deserves. Yet it was this man's ability to crystallize Faraday's ideas, to make them real and to transform them into a practical understanding of just what light is that marks the beginning of the modern era of light science. It was also thanks to this man's work (and a touch of his own genius) that Einstein came up with his theory of relativity. The man was James Clerk Maxwell.

Born in Edinburgh in 1831, Maxwell was soon to show an easy ability with mathematics and a fascination with nature. His family was comfortably off financially – he spent his boyhood at their manor house Glenlair on the country estate at Middlebie in Galloway, and there was able to let his interests, shared by his father, John Clerk Maxwell, run wild. With a boy's enthusiasm, Maxwell revelled in the new age of science and technology. He was delighted by anything technical, often helping a family friend, Hugh Blackburn, a professor at Glasgow University, with his hot air balloons. But it was colour, the beauty and variety of colour, especially in crystals when they were stressed and distorted (he described them as 'gorgeous entanglements of colour') that drew him into the study of science.

The good times did not continue for very long. Maxwell's mother, Frances, died of cancer when he was eight. After a time with a private tutor, he was sent to Edinburgh to attend the Edinburgh Academy. He was a small boy, more interested in learning than games; he had a stutter and a broad country accent. These added up to make him classic bully-fodder. He was landed with the nickname 'Dafty', which stuck with him for years. But at least in the holidays he could return to Glenlair, to the familiar countryside where he was able to show an interest in the world without being mocked.

Maxwell was sufficiently advanced to transfer to Edinburgh University at the age of 16, but three years later he moved on to the more physics-oriented Cambridge. After a term at Peterhouse College, he moved to Trinity, Newton's old haunt. This move seems to have been inspired by the need to get him under the influence of an appropriate tutor. His friend Peter Tait commented that he had 'a mass of knowledge which was really immense for so young a man, but in a state of disorder appalling to his methodical private tutor'. Professor James Forbes of Edinburgh University remarked to the master of Trinity: 'he is not a little uncouth in his manners, but withal one of the most original young men I have ever met with'.

A most original young man

From the beginning, Maxwell was inspired by Faraday. When he graduated from Cambridge in 1854, he wrote to his mentor and fellow Scot, William Thomson, to say that he intended to attack electricity, beginning with a study of Faraday's work. Maxwell soon expanded his interests wider. Like other truly great scientists, Maxwell's genius lay in his ability to go beyond the narrow understanding of his own field to pull in ideas and associations from elsewhere. He liked to take a pictorial view of what was happening, drawing visual analogies with other aspects of physics and managing remarkably to treat maths in the same visual way – to Maxwell, there was an almost solid link between abstract equations and the real physical world. This shines out in a comment he made when looking back over his career in 1873:

> I always regarded mathematics as the method of obtaining the best shapes and dimensions of things; and this meant not only the most useful and economical, but chiefly the most harmonious and the most beautiful.

It was by analogy with other branches of physics that he was able to pull off what no one had managed before – a description of how light worked. As with so many of the best discoveries, it was at least in part an accident. In thinking about electricity and magnetism and the way they interacted, he played around with similarities between these invisible forces and the way fluids travel through pipes. Treating the invisible ether like a fluid, he was able to link together aspects of electricity and magnetism to produce an unexpected result. He found that an electrical wave and a magnetic wave could support each other as they travelled through the ether – but only if they moved at one specific speed. When Maxwell calculated that speed he was amazed to discover that it was exactly the speed of light.

Spurred on by Faraday's speculations, Maxwell made the

bold assumption that light was in fact this interplay of magnetic and electrical waves. He remarked on the speed his calculations forced on the interacting pair:

> This velocity is so nearly that of light, that it seems we have strong reason to believe that light itself (including radiant heat and other radiations if any) is an electromagnetic disturbance in the form of waves propagated through the electromagnetic field according to electromagnetic laws.

Leaving behind his mechanical fluid analogy, Maxwell went on to push the theory fully into the known behaviour of electricity and magnetism until he was able to assemble eight groups of equations that described the workings of these electromagnetic waves – the innermost secret of light. These works of mathematical genius were later simplified by two other physicists, Oliver Heaviside and Heinrich Hertz, into four neat equations that would join a handful of others in forming the description of 'how everything works'. We'll come back to the equations in a moment, but first let's explore the simplicity and power of Maxwell's accidental description of light.

By its own bootstraps

According to Maxwell's picture, light is a balancing act, a constant self-creating marvel. Electricity, when moving along, generates magnetism. Magnetism in motion generates electricity. Faraday had demonstrated both these facts. Light was the result of the interplay of the two at just the right speed – a ripple of electricity that supported a ripple of magnetism that itself supported the ripple of electricity. It was a perfect, self-sustaining perpetual motion machine.

This ability of light to haul itself along by its own boot-straps depended on movement. Unless light maintained its own, definitive speed, the electricity would not generate the right amount of magnetism, the magnetism would not produce the appropriate electrical current and the whole delicate balance would collapse. It was this speed that had enabled Maxwell to make the leap from describing a theoretical interaction between electricity and magnetism to understanding the mechanism that underlies light. Maxwell's picture would work *only* if the waves moved at this particular speed, and it was just too much of a coincidence that it happened to be the speed of light.

The detailed workings of Maxwell's equations are about maths rather than light, but the equations themselves in their final form are very stark and simple. In that starkness is a kind of beauty. If the thought of equations turns you off, don't try to think of it as maths, just consider how this compact set of shapes can open up the secrets of light itself:

$$\nabla \times \mathbf{E} = -\frac{\partial}{\partial t}\mathbf{B}$$

$$\nabla \times \mathbf{U} = -\frac{\partial}{\partial t}\mathbf{D} + \mathbf{J}$$

$$\nabla \cdot \mathbf{D} = \rho$$

$$\nabla \cdot \mathbf{B} = 0$$

The formulation looks odd because these are equations dealing with more than one dimension. The point-down triangle, called 'del', is a way of describing change in all three dimensions at once. But the meaning of the equations is quite simple.

The first is a reworking of Faraday's law, showing how a changing magnetic field generates electricity. The second describes the way an electric current will generate magnetism. The third provides a direct link between the electrical field that is generated and the electrical charge, the sum or absence of electrons. And finally the last equation, which explains why magnets always come with both north and south poles, says that there are no isolated magnetic poles.

Defining colour

Although Maxwell's triumphant revelation of the constant dance of electricity and magnetism at the heart of light was his greatest contribution to its story, his early fascination with colour was to bring out other unexpected details.

His biggest contribution here was to dismiss one of Newton's minor errors. Newton, though well aware that mixing colours of light wasn't the same as mixing pigments in paint, had claimed that green could be produced by a mixture of yellow and blue lights. Maxwell picked up on his old Edinburgh professor James Forbes' observation that blue and yellow never made green, but instead 'a yellow-grey or citrine'. Fixing instead on the primary colours of red, blue and green that Thomas Young had first proposed, Maxwell was able to put together the first clear picture of how the eye perceived colour, with separate components picking up each of these three primaries – and hence to explain the way colour blindness works when one of these facilities breaks down. As before, it was Maxwell's ability to meld maths and experimental results that made his work a triumph, describing for the first time in a mathematical form the way the three primary colours combine to provide any hue. His approach is still used today in producing colour on computer screens and TV sets.

Maxwell also took an interest in photography, then solely a black and white affair, and was determined to bring his fascination with colour into the field. Although colour photography would not become commonplace for another 100 years, it was as early as 1861 that Maxwell managed to produce the first true colour photograph. His process required a long exposure, so his photographer, Thomas Sutton, opted for a subject that was not going to a move and one that reminded the world of Maxwell's origins – a piece of Scottish tartan ribbon.

That Maxwell achieved a full colour photograph was more fluke than careful science. Unknown to him, his colour plate was not sensitive to red, so the tartan should have come out in a pasty combination of blues and yellows and greens. As it happened, though, the chemicals he and Sutton used were sensitive to ultraviolet, which happened to be produced most strongly from the red dyes in the tartan. The ultraviolet image was coloured red in the final picture, producing a result that looked like the original entirely by accident.

Chapter Seven

Death of the ether

More light! Give me more light!
JOHANN WOLFGANG VON GOETHE

*We shouldn't be so provincial: what we can
detect directly with our own instrument, the
eye, isn't the only thing in the world!*
RICHARD FEYNMAN

Maxwell's amazing revelation of light's electromagnetic nature prepared the ground for many of the twentieth century's most fundamental scientific discoveries. But despite his insight, Maxwell made one wildly inaccurate assumption about light. Towards the end of the 1800s, quite unintentionally, the American Albert Michelson would discover the truth. Before then, the nineteenth century flair for invention would have brought many new discoveries to the world of light.

Maxwell defined light by the forces that produce it, rather than the eye's ability to detect it. This move away from a dependence on the eye was echoed by new 'colours' that were being discovered, extending the spectrum beyond the visible limits of red at one end and violet at the other. The light that our eyes respond to occupies a tiny segment near the middle of this huge span of colours that we can never see. If the whole range of light, visible and invisible, were represented by a rainbow, we can see only a thin slice out of the green segment. Some definitions of light only apply the word to the visible spectrum, but 'light' is a lot less clumsy than 'electromagnetic radiation', so let's stick to 'light' for both visible and invisible rays.

The King's astronomer

The first sign that invisible light could exist predates Faraday and Maxwell. It was back at the start of the nineteenth century that astronomer William Herschel made a surprising discovery.

Herschel had an unusual background for a scientist. Born in Hanover in 1738, Friedrich Wilhelm Herschel was the son of the bandmaster to the Hanoverian Guard. Not surprisingly, he developed an early interest in music. Herschel joined his father's band at the age of 14, but military bandsmen of the time could not expect to stay comfortably in the barracks for too long. Four years later he was sent over to England as part of a national defence force in case of French invasion (George II was King of both England and Hanover).

When Herschel returned to Hanover, he soon applied for a discharge, which was duly granted. For some reason Herschel has been described as deserting, but there is no evidence to suggest this. Once untangled from the military regime he was eager to get back to music. His time in England had been pleasant; he had picked up a good smattering of English there, and so he joined his brother Jacob in a trip over to London in 1757. It wasn't intended to be a permanent move, but Herschel could hardly have predicted what was to happen to him.

Before long, Herschel's skills at the organ won him a post in Halifax in the north of England, but the job was not highly paid and he remained on the lookout for something more financially rewarding. A position came up to play the organ at the Octagon Chapel in Bath, a city that was then at the height of its popularity with the fashionable set. Herschel's talent and cosmopolitan social skills won him the job. When he wasn't playing, he took on private pupils and composed. As a successful musician in the most exclusive resort in the country, Herschel was not short of money, and increasingly had spare time in which to amuse himself. He took up astronomy.

For many wealthy people of the period, astronomy was a casual interest, and this seems to have been the case originally with Herschel when he hired a small telescope and made occasional attempts to scan the heavens, but his real interest was sparked by the thought of making his own telescope, about the time he moved to a larger house in Bath's elegant New King Street. Herschel had no experience of instrument making, but he had enthusiastic helpers in his sister Caroline (who was by now his caretaker) and his brother Alexander. The enthusiasm was needed. It was very easy to fail when every aspect of production, including polishing and shaping a metal plate to a perfect mirror surface, was a matter of trial and error. But by 1774 he had constructed his own telescope, a five foot long tube with an eight inch wide mirror at the end.

His greatest astronomical achievement was realized with one of the series of small telescopes he made in the next house he bought, a little further down New King Street. He discovered what he thought was a new comet, but in fact proved to be an undiscovered planet, Uranus. By now astronomy was an obsession. Frustrated by the limitations of his telescopes, Herschel was determined to go further than anyone had before, and built a giant device with a mirror three feet across in a tube a good 30 feet long.

Just producing the mirror was a hazardous task. In the confined cellars of his house, molten metal was poured into a mould made of dried horse manure. By now Herschel had workmen to help, which was just as well. Before the mirror had solidified the mould split open and the fiery metal poured across the flagstone floor. The stones, stressed intolerably by the extreme heat, cracked apart and pieces of them flew across the room like shrapnel, leaving the poor workmen to run for the door as the shards of stone flew in all directions and ricocheted off the ceiling.

Herschel's next attempt at producing the great mirror succeeded. Soon his fame was in danger of eclipsing that of the Astronomer Royal. The King, George III, now better remembered for his descent into madness than his patronage, was an enthusiast for anything scientific, with a particular interest in astronomy. He was sufficiently impressed by Herschel's work to devise a special post for him – King's Astronomer. At last Herschel could give up music and dedicate himself full time to the skies. But there was a price to pay. Herschel was too far from the court in his comfortable house in Bath. He would have to move to a more convenient location. He settled on Slough.

This grimy industrial town to the west of London seems an unlikely choice for an observatory site now, but at the time it was little more than a village, close enough to the capital to be accessible in a day, but far enough out in the countryside to avoid any distortion from smoke, heat and light, and handy for the royal residence at nearby Windsor. In Slough, Herschel built his greatest telescope ever, a huge monster with a 49 inch mirror and a tube forty feet long that was mounted in a great wooden structure of poles and ladders, allowing it be tilted and turned to take in any point in the heavens.

The theory was good, but the great telescope was difficult to manoeuvre and used an odd optical system that had no small mirror, but instead tilted the main mirror at an angle so it focused at the side of tube. Herschel's design made it difficult to see a clear image and the telescope was never to produce as exciting a result as the small instruments he first built. This didn't matter for Herschel's one significant contribution to the history of light, which did not even involve a telescope.

Heat from light

Herschel's home and workplace in Slough, Observatory House, was more than a site for telescopes, it was a full-scale scientific laboratory. In 1800, the year before Young presented his paper on light and colour, Herschel was playing with light as Isaac Newton had before him, letting a thin slice of the Sun's rays fall on a prism. Herschel was investigating the heating effects of the spectrum of colours. That sunlight was warming had been obvious since ancient days. More recently, with the invention of the thermometer, it had been noticed that this wasn't merely a subjective observation. Thermometers placed in sunlight rose in temperature as the light heated them up.

Herschel was interested in the way that the colours in light's spectrum behaved differently (though he was unaware of Young's assertion that the colour variation was due to differing wavelengths). Instead of just leaving a thermometer in sunlight, he tried it out in different parts of the spectrum. Despite his enthusiasm for bigger and better telescopes, Herschel's equipment was very simple. He used a conventional prism to project a spectrum onto a screen that was held in a moveable frame not unlike the mount of a shaving mirror. Down the centre of the screen was cut a slit, and through this, a narrow portion of the spectrum continued to fall on a thermometer. As he moved the slit to take in different parts of the spectrum, Herschel found that there was a corresponding change in temperature. There was more heat towards the red end and less towards the blue.

A lifetime's studying the stars had left Herschel very careful about positioning. Pinning down the precise location of an object in the sky took exacting measurement. When he came to try out the spectrum, he wanted to be very sure just where the red light ended. Not trusting his eyes, he continued to take measurements, moving the slit further and further into the blank space beyond the red. To his amazement, the readings carried on a long way, getting stronger and stronger. There was no sudden break at the end of the visible red light – instead light continued in an invisible form with an even stronger heating ability. He called this extension of the spectrum infrared.

Herschel returned to astronomical researches until his death in 1822, but as soon as he had published the results of his discovery of infrared, another scientist was spurred on to explore the opposite extreme of spectrum. Why, after all, if the bottom end doesn't stop at red, should the top finish with violet? The man in question was the German Johann Ritter, and to make his discovery possible he took the first steps in the development of photography.

Beyond the violet

The basic concept behind photography is now thought to go back a long way, perhaps even as far as the thirteenth century. One of the early influences on Roger Bacon (see page 36) was a Bavarian priest who spent much of his time teaching at the University of Paris, Albertus Magnus (later Saint Albertus). Just as Bacon became known as Doctor Mirabilis, Albertus was given the title Doctor Universalis, the universal doctor, because of his interest in all of nature. It was he that was largely responsible for the recovery of the ancient Greek texts on science from their Arabic translations, an interest in the exotic that gained him a reputation as an alchemist and magician that he never managed to shake off. Many of the legends later attributed to Bacon, including the talking brass head, were originally stories about Albertus.

Some sources suggest that Albertus was the first to notice that chemicals with silver in them turned black when exposed to light (though this has also been attributed to two seventeenth century characters, mineralogist Georg Fabricus and chemist Angelo Salo). What is certain is that in 1727 Johann Heinrich Schulze found that a particular silver salt, silver nitrate, would retain a darkened image when a picture was projected onto it using a stencil. Although it would be another hundred years before a true photograph would be made (see page 164), Johann Ritter found a use for this crude technique.

In 1801 Ritter read about Herschel's discovery and realized that it would be difficult to use thermometers to detect invisible light up beyond violet where light's heating effects were getting less and less. Instead, he hoped that the change in colour of silver nitrate would be triggered by all light, not just the visible part of the spectrum. His guess paid off. The paper soaked in silver nitrate was darkened by a slice of the spectrum well above the visible portion. Another piece had been added to the jigsaw of light – ultraviolet – and all this over sixty years before Maxwell established for certain just what light was.

Maxwell's heritage

Maxwell's work provided the spur to go further still in exploring the spectrum. Heinrich Hertz, professor of physics at the technical school in Karlsruhe, Germany, was impressed by Maxwell's arguments and wanted to find an experiment that would demonstrate the reality of Maxwell's equations. In 1888 he built a simple apparatus to demonstrate the strange, electromagnetic waves in action. It looked like a prop from a Frankenstein movie. Perched on a wooden stand was a brass rod around 30 centimetres long with a small gap in it. By putting a high voltage oscillating current across the gap he produced a spark.

To complete the experiment, Hertz blacked out the room until it was completely dark. His spark device, buzzing and flashing with electrical power, must have looked quite fearsome. At the far side of the room was another, similar rod with a narrower gap. At first the brightness of the main spark half-blinded Hertz, but gradually his eyes got used to the darkness. He saw what he had hoped for, but could never be sure would be there. Across the second gap was a faint glow, the result of tiny sparks jumping between the sections of rod.

An electromagnetic wave, a form of light, was crossing the gap and setting up the second spark, but what was this wave? No visible light linked the two ends of the apparatus. From the length of the brass rods and the frequency of oscillation of his electrical current, Hertz could work out where this new form of light fitted into the spectrum – and found that it sat well below the infrared. Hertz thought his experiment was a useful demonstration, and valuable for increasing the understanding of light, but he had no interest in searching for practical applications for his 'electric waves'. It took the young Italian inventor Guglielmo Marconi to read about the invisible waves and think that it could be possible to send telegraph signals this way without any wires in between the telegraph stations. Hertz's simple demonstration became Marconi's wireless telegraphy – radio – the basis of all our broadcast communications.

X-Strahlen

The same year that Marconi built his first successful radio equipment, 1895, a German scientist accidentally found himself experimenting at the opposite end of the light spectrum. Wilhelm Conrad Röntgen was not intending to work with light at all. He was investigating a mysterious device called a cathode ray tube.

The British scientist William Crookes had been experimenting with small-scale lightning, sending electrical sparks through the air between two metal spikes. He wanted to try out his miniature lightning bolts at different atmospheric pressures, just as would happen at different altitudes in the sky. He sealed up the spikes in a glass tube and pumped out some of the air. But as the pressure reduced, the sparks disappeared. Instead, a strange, unnerving glow filled the container. When he reduced the pressure even further, the disembodied radiance disappeared – instead the glass itself began to glow.

Crookes was fascinated by the mysterious rays that caused this glow, called cathode rays because they seemed to originate from the negative electrical point, the cathode. He showed that they could cast a shadow when hitting a metal plate, and that they could be diverted using a magnet, shifting the pattern on the glass or special fluorescent material. (This ability of a magnet to displace cathode rays is the method that was used to generate images on all TV sets and computer monitors until LCDs and plasma screens came along.) Crookes was convinced that cathode rays, despite their name, were actually a stream of negatively charged particles, though many of his contemporaries thought they were a type of electromagnetic ray, related to light. Crookes had also noticed that photographic plates he kept near his cathode ray equipment would get 'fogged', darkened until useless – he even sent back one batch of plates to the manufacturer complaining that they were faulty.

Röntgen had only just been born when the cathode ray tube was first made. He spent his first three years in Germany until 1848 when his family moved to Holland, and with the rest of them he was made a Dutch citizen. Reaching adulthood, he wandered through a number of European universities, before settling down at Würzburg, where he was awarded the post of Professor of Physics. Like many of his contemporaries, he was fascinated by the ghostly behaviour of cathode rays.

One afternoon Röntgen was experimenting with a Crookes tube. The tube was surrounded by black cardboard to cut out the glow from the glass. Röntgen was using a screen coated in barium platinocyanide, outside the tube. On this occasion, when he switched the tube on, the screen was standing alongside the experiment rather than at the target end. To Röntgen's surprise, the screen began to glow.

This shouldn't have happened. The stream of electrons that made up the cathode ray was heading out of the front of the tube. It seemed that when it hit the target, something flew off sideways, something strong enough to penetrate the black cardboard. After eliminating the possibility that the cathode ray itself was escaping (cathode rays, being a stream of electrical particles can be deflected with magnets – these rays could not) it became increasingly obvious that he had discovered something new. A ray that could pass through objects that stopped ordinary light dead.

When Röntgen presented his discovery to the less than prestigious gathering of the Physico-Medical Society of Würzburg he referred to the new breed of light as X-rays (*X-Strahlen* in German), using X to denote something unknown and mysterious. His discovery was later given the official title of Röntgen rays, but it was already too late. Röntgen's much more evocative name had stuck, and X-rays they remained. For his discovery, Röntgen won the 1901 Nobel Prize for physics, the first ever winner of this honour.

So dramatic was the behaviour of these X-rays that Röntgen could not really believe that they were a form of light. In his paper he points out various ways that light acts, but X-rays seem not to. Hunting around for something to explain this 'light that was not light' he hit on an interesting if totally incorrect thought. Light was thought to be a side-to-side (transverse) wave in the ether. So what if this new discovery, the X-ray, passed through the ether as sound does through air, by squashing and relaxing the ether – a longitudinal wave as Röntgen called it? It was a nice idea, but one that really pre-dated Maxwell's proof that light was electromagnetic.

In practice, Röntgen's problems in identifying X-rays as the next step up the spectrum from ultraviolet were simply caused by the power and the tiny wavelength of the X-ray. It wasn't long before his objections were overcome and the X-ray's place in the family of light was firmly established. But the aspect of Röntgen's paper that immediately captured press attention was a single photograph. He had shone the X-ray through his wife's hand. Falling onto a photographic plate afterwards, the ray produced a picture of her bones, a human skeleton seen for the first time while still inside the body. The medical applications were immediately apparent, but the popularity of this technology was strongly linked to the novelty of this 'X-ray vision'. Well into the twentieth century, hideously dangerous do-it-yourself X-ray kits to amuse friends and relations were popular features of amateur electrical magazines.

The full spectrum

Such amateur involvement was never an option with the next step up the electromagnetic spectrum. Instead it was a by-product of Ernest Rutherford's investigations into the nature of matter itself. Rutherford was a New Zealander, but made his greatest discoveries at the University of Montreal in Canada, where he worked from 1898 to 1907, and subsequently in England, first at Manchester and then at Cambridge University. It was Rutherford who was to give us the picture of the atom that is still used today – a dense nucleus with a positive charge, surrounded by a cloud of negatively charged electrons. One of the key phenomena that led to this understanding was that of radioactivity.

The existence of radioactivity had been discovered by Frenchman Antoine Henri Becquerel, just after Röntgen tripped over X-rays in 1896. Like Röntgen's, Becquerel's find was an accident. He had left some salts of the element uranium on a covered photographic plate and found that an area of the plate below the salts became blackened. The uranium was spontaneously giving off energy, a phenomenon that was later given the name of radioactivity. It was Rutherford, though, who found that the radioactive output could be split into two varieties, which he labelled alpha and beta.

Rutherford showed that the beta ray was in fact a stream of electrons, identical to a cathode ray, though much more powerful. Like the cathode ray, the electrons could be made to shift off course given a suitably strong magnet. Initially the alpha ray showed no sign of bending, but eventually, with a stronger magnet still, it too was made to bend, in the opposite direction. Both rays were renamed particles. The positively charged alpha particle was later found to be the nucleus of a helium atom. With the two types of particle deflected out of the way, Rutherford found that a third stream remained, gamma rays. These, like X-rays, stubbornly refused to be influenced by electricity and magnetism. Here was another, even more powerful, higher frequency entry in the electromagnetic line-up.

The full range of light's spectrum (Figure 7.1) is now known to extend from very low frequency radio, with wavelengths of thousands of metres, to cosmic rays, ultra powerful

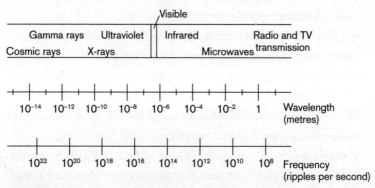

Figure 7.1 Light's extended spectrum. Frequencies are measured as 10^n, where n is the number of zeros after a 1, so 10^6 is 1,000,000. Wavelengths are measured as 10^{-n}, which is the same as $1/10^n$, so 10^{-6} is $1/1,000,000$

gamma rays that hit the Earth from beyond our galaxy, with wavelengths around 100,000,000,000,000th of a metre. It's all the same phenomenon. There is no break point, no distinction in this continuous spectrum. The labels we apply are purely arbitrary. There happens to be a thin slice of the spectrum (with wavelengths from around 1,000,000th of a metre) that influences the eye, the band we call visible light, but the distinction is purely a function of the eye's responsiveness – there is no difference in the light apart from its wavelength or frequency.

Tarnished silver

If invisible light was to spark a whole new communication technology in radio, television and mobile phones, a parallel development would keep invention bubbling away in response to the visual spectrum. Ritter's use of paper soaked in silver nitrate when searching for ultraviolet gave a hint of what was to come. The next year, in 1802, before Michael Faraday was even born, Humphry Davy realized the potential of making pictures using light-darkened silver.

With Thomas Wedgwood, son of Josiah, the founder of the pottery dynasty, Davy had been experimenting with the effect of light on silver compounds. If it seems unlikely that a potter would be involved in such scientific experiments, it's worth remembering that Josiah and Thomas Wedgwood were both enthusiastic amateur scientists – and that finding ways to reproduce a picture that didn't have to be hand painted would be of great interest to a pottery manufacturer. Wedgwood and Davy managed to get some impressive pictures on paper and leather, but could not solve one big problem. The pictures disappeared rapidly unless kept in total darkness, as the same light that made it possible to see the results turned the untouched silver compounds to black, wiping out the image.

It took another son of a famous father to get round this problem. John Herschel had been born in Observatory House in Slough. His environment from the earliest age was saturated with science – it was in his blood. He was a competent astronomer in his own right, taking a telescope all the way down to the Cape of Good Hope on the extreme tip of the African continent to observe the southern skies. Photography would eventually become an essential part of astronomy, but Herschel was interested by the intellectual challenge of overcoming the frustrating blackening of silver-based images. After experimenting with a range of different solutions he found that washing the picture in a particular chemical, sodium hyposulphite, would remove any silver that had yet to react to the light and leave the white parts of the image white for good. This ability to fix the image so that it didn't immediately fade away made photography practical – though it wouldn't be called photography until 1839 when Herschel came up with the name (he also coined the term 'negative' for an image with the blacks and whites reversed).

Herschel might have reduced the technical problems that lay in the way of photography, but he never took the obvious final step of producing a photograph. Such was the frenzy of attempts to make these self-painting images work that it's hard to be sure who was the first, but the oldest surviving example is the work of the French physicist Joseph Niépce, taken in 1822. He used a portable camera obscura to project the image of a landscape onto a sheet of pewter. A layer of tarry 'bitumen of Judea' was applied to the metal. When exposed to light, it sets hard, while the unexposed regions remain soft and easy to remove with a suitable solvent. The plate had to be exposed to the light for a total of eight hours before an image could be produced, but the photographic revolution could truly be said to be under way.

Painting with light

Niépce's interest was theoretical, but a friend of his had a much more practical approach to this new science. Louis-Jacques-Mandé Daguerre was an artist who began his working life painting scenery for the opera. This large-scale experience gave him just the approach that was necessary for the sweeping canvasses that were increasingly popular in the nineteenth century. He began painting huge panoramas, packing vast amounts of detail into his work. Daguerre was fascinated by Niépce's blend of science and art. In particular, unlike the less worldly Niépce, he was immediately struck by the commercial possibilities of such easily produced artwork. He began to spend as much time as he could afford with the scientist, improving the practicality of the photographic process.

Soon after, Niépce's death, Daguerre perfected a process that made a lasting image, which he modestly called the daguerreotype. His big advance on Niépce's approach, was to replace the bitumen with silver iodide on a copper plate, which could be exposed in as a little as half an hour. When the exposure was finished, the silver iodide was pushed into changing colour using mercury vapour (a very dangerous substance that ruined the health of many early photographers), then the image was fixed with a salt solution.

Daguerre's technique was a commercial triumph, though like Galileo before him he was ruthless in the face of competition. Hippolyte Bayard had first exhibited his process before Daguerre, and it was proving a real threat to Daguerre's commercial success. Bayard lost his advantage thanks to an intervention by François Arago, the scientist who proved Fresnel's theories on diffraction right by experiment. While Arago demonstrated admirable independence by standing up for Fresnel, his part in the history of photography is decidedly less respectable. He persuaded Bayard to keep his process secret, giving his friend Daguerre the chance to sell his own technology to the French Academy with no competition.

As a painter of panoramas, Daguerre's natural subjects were landscapes. The first know human portrait, the application of photography that was to turn it from a specialist art form into a mass market, was made in America. Samuel Finley Breese Morse, the inventor of the electric telegraph, met Daguerre on a visit to Paris in 1838. He was fascinated by Daguerre's pictures and began making his own photographs on his return to New York. Before he strayed into scientific experimenting, Morse had been a very successful portrait painter and sculptor. It seems only natural that he would want to make a photograph of a human subject. The following year, with his assistant John Draper, he finally succeeded. It's not the most natural pose, as the sitter had to stay fixed in position for half an hour, and had his face covered with a layer of flour to make his skin brighter, but it remains a ground-breaking exposure.

Negatives and rolls

Two steps remained along the road to popular commercial photography. The first came soon after Daguerre set up business, but in a setting that couldn't have been more different from Daguerre's bustling Paris studio. Englishman William Henry Fox Talbot was brought up in the rural Wiltshire village of Lacock. This beautiful setting, on meadows running down to the Avon river, had been chosen six hundred years before by Ela, Countess of Salisbury as a site for a new abbey. The remains of the fifteenth century buildings were sold off to make a private house in the unholy scramble for cheap property that followed Henry VIII's dissolution of the monasteries. This delightful, rambling house had been Fox Talbot's home and solitary playground throughout his childhood. Now, as an adult, it still fascinated him.

Fox Talbot owned a popular toy of the time, a camera obscura that projected an image onto paper so that a sketch could be made directly over the image. He had heard of Davy's experiments and possibly Niépce's work, and he wondered if he could use something similar to his sketching device to freeze an image in silver. Fox Talbot called his technique photogenic drawing. Although the working process came after the daguerreotype it was not derived from it – in fact Fox Talbot published details of photogenic drawing eight months before Daguerre went public on his invention. Fox Talbot's method has one big difference from the daguerreotype that was to prove invaluable in making photography practical for all – instead of producing a positive image, like the original scene, it produced a ghostly inversion with blacks for whites and whites for blacks.

What might at first seem an irritating drawback proved to be a fundamental advantage. By shining light through the negative image onto a piece of paper treated with photographic chemicals, a positive picture is produced – and once the photographer has the negative, any number of positive prints can be made, while the daguerreotype process produced a single, fixed image at the time of exposure.

The final necessity to ease photography into popular hands did not take place for another fifty years. The need to handle and process individual, specially coated plates, whether metal, glass or paper, made taking photographs a messy business only suited to the professional. In 1884, George Eastman, who was to form the Kodak Company, produced the first roll film – a long sheet of photographic negative that could be moved along, piece by piece, inside the camera and developed later when it held a string of images.

Dissecting motion

Eastman's roll film made another, quite unexpected, development of light-based technology possible. The celluloid sheet of photographic negative material was intended to allow the cameraman to take a series of shots without the complications of changing plates. But what if that series was exposed in quick succession? The result would be to capture motion as a series of images, a dissection of the world of movement into slices on the celluloid roll.

The idea of recording motion in a sequence of photographs was not new, thanks to a long-standing argument in Californian racing circles. In human beings there are only two normal modes of locomotion – walking, where one foot is always on the floor, and running, where both feet leave the floor simultaneously for part of the time. What hadn't been established was exactly what is happening in the more complex field of horse locomotion. It was known that a horse 'flew' when cantering and galloping – the pause was clearly audible in the beat of the hooves – but no one was sure whether horses managed to get all four of their feet off the floor simultaneously at the trot. A railroad magnate, onetime governor of California and racehorse breeder, Leland Stanford, was determined to settle this point, according to legend not only out of curiosity but because he had a significant bet at stake. All the evidence is that the bet, for $25,000, never existed – but it made a good story for the newspapers of the time.

Stanford approached the wonderfully eccentric photographer Eadweard Muybridge (originally Edward Muggeridge) who had moved out to California from the less exotic environs of Kingston upon Thames in the 1850s. Stanford gave Muybridge a huge $40,000 grant to catch the horse, Occident, in mid-trot. The photograph would have to be clear enough to establish whether or not its feet did leave the ground simultaneously. Muybridge arranged a series of cameras along a short racecourse, using threads broken by the horses, or electrical contacts tripped by the wheels of a carriage, to trigger rubber-band powered shutters. The experiment was a success. From the sequence of pictures it became obvious that the horse actually did 'fly' for part of the time. But it also became obvious to Muybridge that he had stumbled on a superb money-spinner.

The flamboyant photographer set off across the Atlantic to begin a European lecture tour, displaying his images on the big screen with magic lanterns. The crowds flocked in to hear Muybridge's story and see the movements of the horse, frozen in time. He was feted at venues like the Royal Institution and the Royal Academy. This shower of glory did not go down too well with Stanford, who produced a book on the experiments that failed to give Muybridge credit. The pair went through a protracted lawsuit over this book: Stanford's money ensured he won the legal battle, but Muybridge won the historical acclaim. He went on to assemble a vast collection of studies in motion (that these often seemed to involve women with no clothes on, was, Muybridge argued, merely so that human motion could be studied more clearly). More significantly still, he invented a projector called the zoöpraxiscope that could replay the sequences of images as moving pictures. He even constructed the first purpose-built cinema at the World's Columbian Exposition, at Chicago in 1893. His presentations might have been short, lasting only a few seconds as the images were around the circumference of a disc, but these were the first true projected moving pictures.

If a series of pictures are shown one after the other quickly enough, the result is the impression of smooth motion. Muybridge amazed his audience by recreating the movements of life on his screen. Since Victorian times this effect has been ascribed to 'persistence of vision', a blending of after-images to make a moving picture, but as a better understanding of the brain has been developed, this interpretation has been shown to be false.

Projecting onto the brain

It is now clear that visual after-images don't form until around 50 milliseconds after the original picture has ceased to be projected, which isn't quick enough to bridge the gap between frames in a movie. Practical experience from the early days of cinematography showed that the pictures had to be changed around 50 times a second to fool the eye. Early silent movies were shot at 16 frames per second, with each frame shown three times, while sound movies run at 24 frames per second, showing each frame twice. The images are on screen for too short a time for persistence to account for the lack of visible flicker. And persistence of vision can't explain the apparent motion we see on the screen, as even if persistence had time to work, it would result in multiple images building on top of each other, not lifelike movement.

Cinema and television work because of the brain's ability to substitute what it thinks is the right thing to see for the actual visual signal it receives. Inside the brain, different visual sensory 'modules' deal with requirements like motion detection, object and pattern recognition and detail selection. The different modules don't handle a single picture, but rather many different elements of the scene. The retina of the eye contains around 130 million light-sensitive receptors. When a photon penetrates to the back of the retina (the photoreceptors are back-to-front with the sensitive part at the rear, a clumsy arrangement that may well be an accident of evolution) it triggers a chemical reaction. This reaction sends a signal back towards the surface of the retina where input from different receptors is combined before feeding the information through the optic nerve to the brain. This nerve has a lot fewer nerve fibres than there are receptors in the eye, so the signal has already been processed at this stage.

The combined image we 'see' is much more an illusion than it appears. It's a reaction to these complex inputs and a combination of the response of the brain modules that cope with motion, pattern, detail and so forth. The suppression of the flicker between frames of a movie, and the merging of still pictures into motion is a side-effect of the way the various complex systems involved in processing the optical data work together.

A similar example of the brain ignoring the true input it receives is the apparently steady view we have of the world. In reality our eyes spend a lot of time darting about. Whether we are looking at someone else's face or the text in a book, our eyes undertake tiny, fluttering motions called saccades. From the differences in outlook generated by these movements, the detail system in the brain can build up a much more sophisticated image than we could manage if everything was taken in via a static, camera-like gaze. Saccades are very quick – the fastest of all external motions of parts of the body, sweeping through an angle of 10 degrees in as little as 1/100th of a second. But we couldn't cope with such a disrupted outlook, so much of the information is edited out by the brain. This misrepresentation of input is much closer to what is happening when we watch a movie than persistence of vision.

The silver screen

The way the brain transforms repeated, subtly changing, images into movement had already been exploited in the mid-1800s using wheels with a series of drawings pasted on the inside and books that were flicked through with the finger, but first Muybridge's zoöpraxiscope, and soon after devices based on Eastman's film, meant that the pictures could be sourced from real life. With the advent of that flexible roll of negative, the full scale moving picture, the cinema, was made possible.

Like many inventions of the time it is difficult to pin down a single individual responsible for bringing the product into existence. Muybridge first used his zoöpraxiscope in 1879. Thomas Edison came up with a crude approximation to the film-based cine camera in 1891, but it was two brothers, Auguste and Louis Lumière, who took the technology and made it commercial. The earliest surviving moving picture is of workers leaving their factory, and this mundane (in fact, truly boring) subject set the standards for many an early film. It didn't really matter what the film was about – as long as it was moving, that was enough to impress an audience.

The Lumière brothers' movie reached the public in 1895, shown at the Grand Café on the Boulevard des Capuchines in Paris. It was a great success and within days the Lumière company had begun churning out a tide of very short films to keep public demand satisfied. The duration of the movie was determined by the length of Eastman's film rolls, 80 feet, allowing for around one minute of footage on the screen. The brothers called their early camera-come-projector the Cinématographe, from which the British name for the medium 'cinema' is derived. Any developments since have been driven by the way film is used (for instance cutting together different scenes rather than portraying continuous action) and in the underlying technology, such as the move to digital capture and editing, rather than a fundamental change in the part that light takes in the process.

A new lamp

Moving pictures weren't Edison's only involvement in the technology of light. His best known light-based invention, the electric lamp, was a late arrival in the long history of artificial lighting.

Thousands of years before there was any idea of what light might be, its value was abundantly clear. But the obvious natural light sources, the Sun, Moon and stars, had limited availability. There were, however, more portable natural sources of light. There was Galileo's solar sponge (see page 68), glowing with the cold light of radioactivity; the ominous glimmering of bacteria on rotting meat; and the radiance of naturally phosphorescent objects like fireflies, deep sea fishes and some plankton.

This glow of the sea, caused by a myriad tiny one-celled creatures, had been recorded as early as 500 BC. Charles Darwin, crossing the South Atlantic in 1832 on his famous voyage on the *Beagle* remarked:

> The sea from its extreme luminousness presented a wonderful and most beautiful appearance; every part of the water which by day is seen as foam, glowed with a pale light. The vessel drove before her bows two billows of liquid phosphorous, and in her wake was a milky train. As far as the eye reached the crest of every wave was bright; and from the reflected light, the sky just above the horizon was not so utterly dark as the rest of the Heavens.

Such bioluminescence is the result a reaction that turns chemical energy into light, blasting photons from the electrons around the molecules of complex chemicals. It is practically a reversal of the process that provides the oxygen for our air and keeps most plant life flourishing – photosynthesis. The first hint of this natural life-support mechanism was discovered by Joseph Priestley, a Yorkshire-born minister in the Dissenting Church. By the mid-1770s, Priestley was finding it difficult to apply himself to both his ministry and the scientific investigations that had interested him since his college days. He took the opportunity of working for William Petty Fitzmaurice, the second Earl of Shelburne. Shelburne wanted a librarian, but more than that, a man with a sharp intelligence with whom he could discuss literary matters. In return for his company, Shelburne was happy to support Priestley's scientific work.

While with the Earl, Priestley spent much of his time investigating the nature of air. In fact he succeeded in separating off oxygen, though he didn't recognize it as such. Priestley supported the popular theory of the time that there was a component of matter called phlogiston that made it flammable. One of his experiments involved putting a candle under a sealed bell jar. The candle would burn out long before it ran out of wax – Priestley believed the phlogiston was being exhausted. He also discovered that a mouse could equally 'injure' the air. Luckily for at least some mice, he found that it was possible to restore the injured air and the mouse by putting a plant under the jar.

Priestley never got any further with photosynthesis, though he made other chemical discoveries and continued to have a stormy life which itself seemed imbued with more than the usual level of the imaginary phlogiston. Not only was one of his books, *A History of the Corruptions of Christianity* officially burned, but his house was torched by angry mobs when he openly supported the French Revolution. Priestley, who had long since left the Earl of Shelburne over religious differences, emigrated to the United States, where his revolutionary inclinations would be more appreciated.

Only four years after Priestley's discovery, Dutch court physician Jan Ingenhousz managed to take one step further.

Ingenhousz, by then living and working in England, repeated Priestley's experiments, linking the plant's ability to recover the mouse to the energy of the Sun. Unfortunate mice that were put through the experiment in the dark never recovered. It was only when the plant was in sunlight that the mice came back to life.

The final parts of the international effort to understand photosynthesis went to French pastor Jean Senebier and Swiss scientist Theodore de Saussure. Around 20 years after Ingenhousz's work, in 1796, they showed that Priestley's injured air was in fact carbon dioxide, and that plants under the stimulus of light took in carbon dioxide and water to produce oxygen and biochemical carbon chains. Only a tiny part of the Sun's light hitting the Earth – less than one per cent – is used up in photosynthesis, but this powers the whole living structure of the planet. Surprisingly, three-quarters of the energy consumed goes not to trees and grasses but to the tiny specks of algae that float in the seas.

The chemical processes in photosynthesis are complex and often amazingly fast – some of the reactions are the fastest ever measured at under 1/1,000,000,000,000th of a second. The light is absorbed by pushing up the energy of electrons in special colouring materials like chlorophyll. The energy from the light is then transferred in chemical form to an in-plant reactor, the photosynthetic reaction centre, where the fundamental reaction that produces oxygen is performed. Different plants have different levels of oxygen production – for sheer volume, it's actually plankton in the seas that make the greatest contribution, but a high output photosynthetic species like corn can produce enough oxygen to support over 300 people from each hectare of planting.

Fire light

This biological absorption of light works by cold chemical reaction, but natural cold light sources proved inadequate to supplement the Sun for lighting. Despite the example of bioluminescence, for most of history artificial light has been inseparable from flame. Fire not only provides the twin benefits of heating and cooking, it generates light as the heat of the reaction stimulates electrons in the material being burned to give out photons. When it was noticed that the cooking fire made it possible to extend daylight hours into the evening, different materials were tried to make the light source more portable and practical. Oil lamps and candles dominated from biblical times to the mid-1800s. The only real breakthrough was in the introduction of mantles – fine meshes of metal or treated fabric that were heated to white hot by the flame and so gave out a more even, whiter light.

The rule of the oil lamp and candle was first threatened by the easy production of gas. Once it was possible to make gas on demand, this new fuel could provide lighting that was available at the turn of a tap whenever it was required. There was nothing new about the lights themselves – only the fuel had changed. The earliest gas lamps were fearsome contraptions, little more than a flat spray of flame, but with the use of mantles they became more controlled.

Even so, the gas light was a dangerous affair, and Faraday's invention of the generator made an alternative possible. Humphry Davy, Faraday's one-time mentor, was inspired by Faraday's invention, using it to heat a platinum wire until it glowed brightly with the electricity passing through it, but the wire never survived long enough to be the basis for a lamp.

If it was hot enough to provide a light, it was hot enough to melt and burn. Davy also investigated electric arcs – making an electric spark jump across a gap, creating an intense white light. The arc light was commercially successful from the 1860s (it was first used practically in a lighthouse in Dungeness, England in 1862), but it was never a realistic proposition as a replacement for gas or oil lighting, because the exceptionally hot arc was dangerous and required too much maintenance.

With arc lights dismissed as too risky and electric lights using wire filaments impossible to keep alive, it looked for a while as if a third possibility would be the first commercial threat to the gas light. The technology was related to the cathode ray tube. Heinrich Geissler had been experimenting back in the 1850s with a device that used the glow of a sealed, low pressure tube as a source of light. But Geissler tubes proved too weak to be of practical use. Meanwhile, many others tried following in Davy's footsteps, patenting a range of electrical lamps that relied on heating up a wire, so-called incandescent lights. All of these prototype tickets to millionaire status failed. Until the magical year of 1879.

Edison vs. Swan

In 1879, not one but two inventors succeeded in keeping an incandescent light intact. Although his claim to be the first is anything but robust, the name that will forever be linked with the electric light is that of Thomas Alva Edison. Edison was the archetypal American self-made man. Sometimes his lack of education has been exaggerated to emphasize how much he hauled himself up by his own bootstraps – some claim he had as little as three months formal schooling – but it's certainly true that he had very little formal education when compared with most of the contributors to the history of light.

In fact Edison did attend school at Port Huron, Michigan, when the family moved there from smalltown Milan, Ohio, in 1854 so Edison's father Samuel could take a job as a carpenter. Young Edison was seven at the time, his distinctive middle name given in honour of a family hero, Captain Alva Bradley who had a fleet of ships on Lake Erie. Thomas Edison was no great scholar, in part because his hearing was poor and he simply could not keep up with the classroom teaching. When he was 10 his mother decided he was getting no real benefit from the school and took him out to teach him herself. Nancy Edison was a great believer in the educational value of books, and luckily Edison had picked up the rudiments of reading, so she put him on a crash course of learning through the written word.

By the time he was 12, Edison had a job, working as a train boy on the Grand Trunk Railway. There was nothing unusual in being employed at this age – it was a commonplace of the time – but Edison was unique among train boys in not only acting as general gofer on the train, but also having his own on-track laboratory. Ever since Nancy had taken him out of school and introduced him to chemistry books Edison had been experimenting. Before long, his makeshift laboratory had been moved to the basement of the house as he caused too much mess elsewhere, but where most people would find that the move to full-time employment put an end to their youthful experiments, Edison saw it as an opportunity to grow.

He persuaded his employers to let him use an empty freight car as his laboratory and spent all the hours he could experimenting, first with chemical reactions and then with electrical machinery. But Edison's interest was never in pure theory. From the very earliest days he was on the lookout for opportunities for self-improvement. Travelling the railway as he did, he soon spotted an opening for using his mobile base as a communications vehicle. He was still only 15 when he added a small printing press to the laboratory and brought out a weekly newspaper, the Grand Trunk Herald. Then, according to the Edison legend, came a different type of breakthrough.

It's easy to imagine the scene. It was a foul night of swirling rain. Edison was waiting at a station to pick up the train for his latest duty. He saw a boy, the son of the stationmaster, playing by the railside. At the last moment, as the train began to slow for the station, the boy fell onto the track. With no one else near enough to help, Edison jumped onto the track and dragged the boy clear. The thankful father is said to have taught Edison how to use the electric telegraph and so to have opened his way to a new career building sophisticated devices for the telegraph system.

It's a nice *Boy's Own* story, but it's probably nothing more than inspirational fiction. It simply isn't necessary to explain Edison's success. He could hardly have failed to be aware of the telegraph, the information lifeline of the railway, and his fascination with electrical and mechanical gadgets would have made sure that he took a keen interest in the equipment whenever he had the chance. A twelve-year-old boy who could persuade a railway company to give him the use of a freight car as a personal laboratory would have no trouble getting access to telegraphic equipment.

At every opportunity Edison collected scraps of information on technical discoveries and new inventions. In his mobile laboratory he began to make real progress of his own. The first

success was an improved telegraph, which soon raised enough
cash to enable him to move from the freight car to a landside
laboratory in Newark, New Jersey. For the rest of his life he
and his growing team would pour out inventions, some as well
known as the phonograph, others (like the electric pen) that
would remain obscure. But little would have more impact than
the electric light. Edison himself was impressed by the possibilities from the very beginning. When the light bulb went
public in 1879, with characteristic modesty he said:

> We are striking it big in the electric light, better than
> my vivid imagination first conceived. Where this thing
> is going to stop, Lord only knows.

Edison's lamp was a great success. It clearly wasn't anywhere near the first electric light, though it was one of the
earliest to be practical. But only *one* of the earliest. In the same
year as Edison's original development, 1879, but eight months
earlier, the English inventor Sir Joseph Wilson Swan demonstrated his own electric light bulb, based like Edison's on a
carbon wire. Swan, more of a scientist and less of a businessman, hadn't bothered with the level of patent applications that
Edison had. Nor had he the same cutthroat commercial sense.
Edison's reaction to hearing of Swan's earlier invention was to
launch a patent infringement prosecution.

Patent law often seems to favour the commercially strong
rather than the most original thinker, but in this case Swan's
earlier invention was recognized by the court and Edison failed.
As part of the court settlement, Edison was obliged to recognize Swan's independent and earlier invention and to set up
a joint company, the Edison and Swan United Electric Light

Company to exploit the incandescent bulb. It's churlish to suggest Edison doesn't deserve his place in the hall of fame, but Swan rarely gets the recognition he deserves as the true inventor of the practical electric light.

Geissler's glowing tubes might not have had the early success of incandescent bulb, but they still had an important future. Commercial products using various descendants of the tube have been around since the beginning of the twentieth century. Today they're common in street lighting, fluorescent tubes and low-energy light bulbs. All these lights rely on the way that connecting a high voltage discharge across a tube filled with low pressure gasses will produce a glow of light. Some of the electrical energy being poured through the gas is absorbed by the electrons in the gas molecules. Before long the energy is given out again as a photon – a particle of light. Different gases (for example neon, mercury vapour and sodium vapour) produce different colours of light specific to the chemical element, depending on the natural energy levels of their electrons.

Unfortunately, though, these lights have a strong, unnatural colouring. Fluorescent tubes are based on the same technology, but the tube is coated inside with a material that glows itself when the bright light hits it. The colours in which these phosphors glow need not be the same as the original light output by the electrical discharge – in fact much of the light inside a fluorescent tube is in the invisible ultraviolet. The resultant fluorescent glow can be made much closer to the colours of natural light, at the loss of some intensity.

With electric lighting in the ascendant, the gas light companies had every opportunity to get in on the act. Before Edison could sell electric lighting he had to set up a power network to work his bulbs, and he was having difficulty getting them to operate more than around two miles from his generators. Making the classic business error of missing a revolution in the marketplace, the gas lighting companies tried to counter Edison and Swan's launch with bigger and better gaslights. The industry was doomed.

Sky blues

For many years artificial light manufacturers tried to duplicate the exact colours of natural light. It proved surprisingly difficult to manage. But then, sunlight had always guarded its secrets well. As long as humanity has wondered about light, there have been attempts to explain why sunlight turns the sky blue. How could a bright yellow light give the sky such a different and distinctive colour? Elaborate explanations were built about reflections from the blue sea or green grass, but none of them held up to close examination.

A more reasonable possibility was suggested in 1869 by John Tyndall. Tyndall never made it into the big league of scientific names, but he was typical of an age when a scientist had the leisure to explore the whole range of phenomena that interested him. He devised methods of preserving food and studied glaciers. He was a great supporter of Darwin, often raising tempers with his burning enthusiasm. This one-time surveyor on the Irish railways was discovered while a schoolteacher by Michael Faraday. It was Faraday who brought Tyndall to the Royal Institution, where he would eventually succeed his benefactor as Supervisor.

Tyndall knew that the blue tint of the sky was not caused by reflection, that there must be something in the air that made it appear blue. Yet a container of air held against a white background did not show the slightest sign of colouration. Tyndall realized that the air is not made up of pure gases alone; it carries with it billions of tiny particles of dust, thrown up by the storms of the Sahara, blown from seashores and volcanoes, scraped by the wind from every part of the land. Perhaps this dust could be responsible for the blueness.

In his laboratory he filled a glass tube with air and pumped in some very fine dust particles. Then he shone a bright, white light through the tube. The light took on a faint blue tinge when looked at from the side, but was tinted yellow-red where it emerged straight on. Tyndall felt that he was on the right track. It wasn't the dust, he thought, that was coloured blue. Instead, as the light hit the dust it was bouncing off the tiny surfaces of the dust particles. Blue light tended to scatter more easily and so the sky was given a blue tinge when seen at an angle from the source of the light. When the Sun is setting and the light had to pass through a longer stretch of atmosphere, the blue would tend to be scattered even more, leaving stronger reds and yellows in the unscattered direct sunlight.

There was only one problem with Tyndall's theory. It didn't match reality. If dust were the cause of the blueness, you would expect a bluer sky in a dusty atmosphere and a paler, near-white sky when the air was particularly clear. But there was no evidence of especially blue skies over dusty cities like London, and a trip to the top of an Alpine mountain would produce not paler skies, but a sky that was, if anything, even more richly blue.

Tyndall was nearly right, but it took another successor to a famous man to pin down the true culprit. Just as Tyndall had followed the great Faraday at the Royal Institution, James Clerk

Maxwell had handed over the position of Cavendish Professor of Physics at Cambridge to John William Strutt, better known as the third Baron Rayleigh. This title had come, unusually, from his grandmother. Rayleigh's grandfather Joseph Strutt was an eminent politician. Strutt had no intention of giving up a successful career as an MP so he asked that the honour he was being offered for his services to parliament and the army be given instead to his wife.

The third Baron Rayleigh's Nobel Prize (in 1904) was for the discovery of the element argon, but his name is firmly attached to the problem that Tyndall came so close to solving. Rayleigh was inevitably strongly influenced by Maxwell, and took Maxwell's electromagnetic description of light to heart. If light was this combination of electrical and magnetic waves, thought Rayleigh, why should it not interact with the individual molecules of the atmosphere just as much as it did with dust? He envisaged the molecules beginning to vibrate in sympathy with the vibration of the light waves, just as a ball sitting on a tight sheet starts to bounce up and down as ripples are sent through the sheet. This moving molecule, according to Maxwell, would then generate its own new light waves as a result of the way it was vibrating, but in random directions – scattered.

As blue light has a higher frequency than colours like red and green, Rayleigh thought it would shake the molecules faster and generate more new light – hence the predominance of blue in the scattered light. In principle it seems that the sky should appear violet, the highest frequency of visible light, but the Sun's spectrum has less violet than blue in it, and the way the eye's red, green and blue receptors work it is the blue colour that predominates.

Science illuminating art

The varying colours of the sky were as much of a challenge to nineteenth century painters like Joseph Turner as they were to the scientists. Turner's dramatic landscapes make detailed explorations of light, turning the glint of sunlight on the sea or the demonic light of a railway engine in the fog into a sea of light and shade. While the lighting effects in Turner's landscapes parallelled scientific developments, his natural successors, the Impressionists, were directly influenced by it.

The combined work of Young and Maxwell made it possible for the first time to understand exactly how colour addition worked – not the colour mixing of an artist's palette, but the way the mechanism of the eye combines the colours that are seen to make up an image. This understanding of the way red, green and blue light could combine to meet any visual requirement would resurface when colour television was developed. In the older cathode ray TV screens, tiny clusters of red, green and blue elements were used to produce the whole, rich colouring of the picture. But the impressionists got there first.

This production of colour using combinations of individual shades is most obvious in the work of Pointillists, following Georges Seurat, who went out of his way to learn the detail of the optical theories of the day. But even in the earlier, more conventional Impressionists this approach is clear. Seen close up, a Manet face, for instance, is a mess of uneven blobs of colour – perhaps purples and yellows and greens. The paintings simply don't work under close examination. But from a distance, the eye blends the blobs into a smooth, skin colour mix. Traditional painting is 'subtractive', mixing different pigments each of which exclude various colours from reflecting until the resultant colour is left. For such an approach the 'primary' colours are the negative equivalents of red, blue and green – cyan, magenta and yellow, simplified for children to blue, red and yellow, which often causes confusion with the true primary colours. The Impressionist style takes the more natural (yet also more scientific) approach of adding colours rather than subtracting, so that the eye can produce a result.

If the impressionists were the first to truly make use of light in the way they painted, it took the modern abstract form to make light a central part of the art itself. Traditionally art has involved the creation of an object, but from the 1960s onwards, some artists have foregone materials to create works out of pure light and shadow. There has never been a formal movement taking this approach, but it began in Southern California and is sometimes labelled phenomenology. Such artists make use of high technology alongside conventional materials and tend to produce installations rather than art forms – art that you exist within rather than observe from the outside.

The incorporation of actual light into art has sometimes been traced back to the French cleric Father Louis Bertrand Castel, who in 1734 demonstrated his *clavessin oculair*, a clavichord keyboard connected to a series of coloured tapes that moved in front of candles to produce shifting patterns of coloured light. Modern artists using light build complex structures that transform the light, or use optical illusions to confuse the eye. At the popular end of the artistic spectrum it has been holograms and their fascinating ability to freeze a three-dimensional image in space that have most caught the public's eye. As our appreciation of the nature of light has developed, so has the sophistication with which artists have used light in their creations.

Sacrificing the ether

Maxwell had exposed the reality of light as a self-supporting piggyback of electricity and magnetism, but there was still room for some detective work, hunting for the effects of the ether. This tenuous, invisible substance seemed shy to the point of absence. However subtle the search, there wasn't the slightest sign of the ether's existence. Michael Faraday had thought the ether unnecessary for light's electromagnetic progress, but most other scientists accepted that there had to be something for light to travel through. Despite its amazing constitution, light was still a wave, and a wave was a ripple in something – the ether just *had* to be there as the something in which light rippled.

Respected physicists like William Thompson went to elaborate lengths to describe what the ether might be like. Even James Clerk Maxwell, whose equations seemed to make the ether unnecessary, was sure that it was there. He merely thought that he had proved that the electrical and magnetic ethers were the same. He said:

> Whatever the difficulties we may have in forming a consistent idea of the constitution of the ether, there can be no doubt that the interplanetary and interstellar spaces are not empty, but are occupied by a material substance or body, which is certainly the largest and probably the most uniform body of which we have knowledge.

But the ether was a real mystery. How did you prove the existence of something that was so insubstantial that it could penetrate solid glass and so pervasive that it filled all the empty stretches of outer space? It was a challenge that US academic star Albert Michelson found endlessly intriguing. An immigrant as a boy from the German town of Strelno (now Strzelno in Poland), Michelson was obsessed with light. His first attempt to measure its speed was in 1883 and he was still taking measurements with better and better instruments 50 years later.

When it came to detecting the ether, Michelson and a colleague, Edward Morley, had the ideal instrument – the Earth itself. As the planet swept through the ether, a light beam projected from the Earth in the same direction as the planet moved, would be like a fish fighting upstream in a river. The flow of the ether past the light would slow it down, effectively making its path longer. Michelson and Morley devised an experiment to measure this effect. The device was more like a medieval altar than conventional laboratory apparatus. The secrets of the ether were to be sacrificed on this altar of science.

The year was 1887. The laboratory was dominated by a strange, bulky shape. First there was a solid brick base, cemented into place on the laboratory floor. On top of this was a circular metal trough, filled with mercury. In the trough floated a circular construction of wood, nearly touching its sides, and finally on top of the wooden circle was a slab of stone over a metre across. The whole device, seemingly better suited to alchemy than modern science, was designed to protect the slab from any vibration and to keep it moving steadily once started. Such was the lack of friction in the carefully constructed mercury trough that once the slab was rotating at around a turn every six minutes it kept turning for hours. This wasn't Michelson's first attempt. He had been hunting for the effect of the ether for six years, but this time his equipment seemed perfect.

On top of the apparatus, a series of mirrors enabled a beam of light to pass backwards and forwards across the slab. The beam was split into two, with half tracing a path in one direction, the other half travelling at right angles, before the two rays were brought back together. Just as in Young's two slits, these two beams of light then interfered, producing a pattern of fringes that were viewed through a small microscope mounted on the block. Michelson and Morley's device was not designed to come up with any Earth-shattering results. Growing out of Michelson's interest in measurements around light, it was supposed to demonstrate something called the ether wind.

As the Earth moves through the ether it should give light travelling in the same direction a slightly longer path to traverse. Michelson's clever idea was to set up an optical racetrack. The same beam of light, split, was sent off on two paths, each exactly the same length but oriented 90° away from each other. At any one time, one of the beams should be oriented more in the direction the Earth is moving through the ether than the other, so the timings should have changed as the block slowly rotated. This meant that one beam would increasingly have further to travel than the other, and the fringes, viewed through the microscope against a fixed grid, would shift.

In the near-darkness, there was something medieval and mystical about the slowly rotating stone slab. Michelson and Morley had the appearance of acolytes, tending a strange altar of science. But however much they repeated their ceremony, nothing happened. The beams of light threaded their way through the maze of mirrors and arrived back at the microscope. The fringes remained stubbornly in the same place.

However the slab rotated, there was no change. There was no ether wind.

The experiment was repeated over and over again with no better result. With deep reluctance, Michelson was forced to admit that quite accidentally he had shown that the ether did not exist in the first place. Michelson, by now an American citizen, became the first US scientist to win the Nobel Prize for Physics. The ether that had been thought necessary for so long to keep the waves of light in action was dead. Such was the resistance to the results that 13 years later some scientists were still uncomfortable with the Michelson–Morley finding – but it was to hold true, and even more unsettling news was to follow. The fact that light was a wave was a much more solid one than the suggestion that the ether existed. Young's practical work and Maxwell's equations had proved this beyond doubt. And yet a concept bearing a startling resemblance to Newton's particles of light was to surface in the early years of the twentieth century. Traditional science would never recover from the shock.

Fearful symmetry

There was a young lady named Bright,
Whose speed was far faster than light;
She set out one day
In a relative way
And returned on the previous night.
ARTHUR BULLER – *RELATIVITY*

As the twentieth century dawned it seemed that there was little more to find out about light. Experiment after experiment had proved it to be a wave. Faraday and Maxwell had described its makeup as an intertwined magnetic and electrical ripple. Michelson had shown that the ether simply wasn't there. There were just a couple of small details to iron out, little problems where the theory didn't fit what was actually observed, but no one had any doubt that these discrepancies would soon be fixed. They would be – but in a way that would blast apart all that had been assumed before.

Too original a contribution

One of the reluctant revolutionaries was Max Karl Ernst Ludwig Planck. Born in Kiel, Germany in 1858, he was solidly Victorian, feeling the nineteenth century excitement with the benefits of new technology, but also burdened with the nineteenth century assurance that the world as they knew it was *right*.

It was only an urge to make an original contribution to the world that made Planck a physicist. At the end of his time at the Maximilian Gymnasium in Munich he was torn between music and science as a career. He was a superb pianist with perfect pitch and could easily have become a professional musician. It's ironic that having chosen physics as a subject in which he was more likely to make an original contribution, he would spend much of his career denying the implications of his most significant discovery.

Such a future would have seemed unimaginable to the 17-year-old Planck as he started his course at the University of Munich. He was simply looking forward to knowing more; first learning, then applying himself to widening the boundaries of human understanding. Before long, though, he was wondering if abandoning a musical career was a wise decision. He found the physics professor at Munich, Phillip von Jolly, disappointingly weak. According to von Jolly, physics was a complete science and the role of the physicist was polishing what was known rather than opening up new fields. But the quality of the staff wasn't enough to change Planck's opinion of his own chances, so he bolstered his education by reading as widely as he could and by picking up extra tuition at Berlin University. By the young age of 21 he had received his doctorate.

Planck had long been fascinated by heat and energy, and this drew him into the problem that was given the dramatic name of the ultraviolet catastrophe. It was known that every object gives off some light (usually not in the visible spectrum at room temperature, though it becomes obvious if you heat an object up and it begins to glow first red and eventually white hot). Physicists, rather confusingly, refer to this as blackbody radiation, because a pure black object gives out or absorbs radiation perfectly.

From what was observed it seemed that the energy in the light wave was directly linked to its frequency, the rate at which it rippled. The higher the energy, the higher the frequency. This made a lot of sense. When a piece of metal was heated up, as it got hotter and hence more energy was involved, the frequency of light produced went up. But once the maths was worked through, there was a second and more worrying conclusion. The total amount of energy given out also went up sharply with frequency. For ultraviolet light, the energy levels shot off the graph, implying that almost infinite amounts of energy should be pouring out at the highest frequencies. This clearly wasn't happening in the real world – if it had been true, every object, every person would glow with an intense light, rapidly losing all the energy it contained.

It was what Planck later described as a lucky guess that enabled him to get around this ultraviolet catastrophe. He assumed that the energy of the light *was* directly linked to the frequency, but that it was not possible to have light produced at any and every energy level. Instead, a particular atom or molecule producing light was restricted to giving off chunks of energy of a particular size. These chunks or packets Planck called *quanta*, after the Latin term for 'how much'. But if light was made up of little chunks, surely it wasn't a wave, but instead was a particle just as Newton had said so long before? Not according to Planck.

Planck never would accept that light was anything but a wave, seeing his packets of energy as a mere mathematical trick to get the right results. He wrote:

The whole procedure was an act of despair because a theoretical interpretation had to be found at any price, no matter how high that might be.

He felt increasingly out of touch with modern physics, despite winning the Nobel Prize in 1918 for his work on quanta, remarking:

If anybody says he can think about quantum problems without getting giddy, that only shows he has not understood the first thing about them.

Intellectually Planck remained a child of the nineteenth century. His sadness at the direction physics was taken eventually became overtaken by the personal tragedy that dogged the second half of his life. His elder son was killed in the First World War. Both his daughters died in childbirth. And finally his younger son was implicated in a plot against Hitler and executed by the Gestapo. Two years later, Planck himself died.

Destroying assumptions

It was Planck's quantum packets that led to the concept of the photon, but he *knew* that light was a wave – everything that he had been taught proved it. Einstein was less constrained by other people's observations. He was quite happy that light really was a spray of tiny particles. It was the basis for his Nobel Prize-winning work on generating electricity by bombarding metals with light and blasting electrons out. But he found the price that he paid for this belief was a steep one. He had to give up many of the fundamental assumptions that had gone without question in physics for hundreds of years. The new physics that worked at the level of individual particles of light and matter – quantum physics – would test even his ingenuity to the extreme.

Einstein's genius is unquestioned, but his fame outside the scientific community, just like Newton's in his day, is based as much on a legend as the real man. He was superficially affable, happy to discuss his theories with anyone, but held back from real in-depth friendship and rarely managed to work effectively with others. Perhaps this lack of trust in others was as a result of the sad failure of his first marriage, a marriage based on a love that he never again truly felt. But time is being pushed out of place here, a problem that Einstein always seems to engender.

Einstein was born on 14 March 1879 in southern Germany, in the city of Ulm, in a drab block of flats that was destroyed in the Second World War. His father Hermann, from whom Albert would inherit a tendency to be a dreamer, tried earnestly to run businesses funded by his wife Pauline's family – but he seemed unsuited to the aggression needed to make a success of business life. The Einsteins managed to provide a happy home for Albert and his younger sister Maria – always Maja to him – but their finances were never very stable.

Outside the family home, Einstein found things less to his liking. From an early age he hated authority and those who used a position of power to try to control him and his thinking. This reaction that stayed with him his whole life would later emerge in a deep-felt pacifism, but as soon as he began school he found that there was a painful conflict between the learning that fascinated him and the rigid educational system of nineteenth century Germany, which seemed designed to irritate and confine him rather than to expand his imagination.

This distaste for traditional schooling was echoed by the dislike some of his teachers felt for him. By now living in the Bavarian capital Munich, Einstein attended a Catholic primary school (his parents were not practising Jews) where the headmaster once commented that it didn't matter what career young Albert tried, because he would never make a success of anything. At home, playing with Maja in the wild garden of their unkempt house, or more often playing alone in his room, Einstein felt no threat. Here he was the master of his own direction. At the school, where already he was itching to do things his own way, the regime was one of unbending following of rules, an authoritarian discipline that Einstein found galling.

By the time he was in secondary school, at the Luitpold Gymnasium, this resentment was coming through all too clearly. The school put a great stress on a classical education, while Einstein struggled to absorb the dead structures of the classical languages and could not find enough to interest him in the humanities. His lack of enthusiasm got him a reputation of being lazy and uncooperative. To make up for his lack of stimulation from the school, Einstein turned elsewhere. He found a more reliable, more enjoyable source of learning in books, and in the guidance of a young family friend, Max Talmud, a medical student when he first met Einstein. Often invited along to eat with the family, Talmud would feed young Albert's enthusiasm for knowledge in science, bringing along new books and tantalizing titbits of information.

But this comfortable home life was to be pulled apart, leaving Einstein without the anchor he needed to cope with school. Setting off on yet another risky business venture, his father moved the family to Pavia in Italy. Einstein was left behind. His parents thought that this would be the best way to ensure that young Albert's education continued, but the rigid discipline of school now had nothing to cushion it. What's more, before long Einstein would be expected to start his year's compulsory national service. If the school irritated him, the thought of that even more senseless and authoritarian military life was the final straw. Albert decided to follow his family. With no warning, he turned up on their doorstep in Italy.

The whole experience of secondary school, of his short-term isolation and the possibility of national service weighed heavily on the young Einstein. Added to this, his secondary school had expelled him (admittedly after he had already decided to leave). He seems to have blamed his experiences on the German state, feeling that the country's legal structures embodied mindless control. At the age of 16, when most boys are thinking of little else but girls and enjoyment, Einstein made it a personal goal to cease to be a German citizen. His parents were not enthusiastic, but gave in to his relentless pressure and worked through the paperwork to enable him to renounce his German birthright.

There had, though, to be somewhere for Einstein to go. His poor Italian made settling in Pavia an unappealing option. Instead, he targeted Switzerland, acknowledged as the least interfering of all states. In German-speaking Zurich there was an ideal place to start getting a *real* education, the Federal Technology Institute (in German the Eidgenössische Technische Hochschule), universally known as the ETH. While still 16, Einstein took the ETH entrance examination – and failed.

The examination was comprehensive, covering far more than science and maths; Einstein was let down by his highly limited areas of expertise. He was also much younger than the majority of applicants. But the principal of the ETH was impressed with Einstein and recommended that he reapplied after a year at a Swiss secondary school. The tactic paid off. With support from his hosts in Switzerland, the Wintler family, Einstein worked hard for his re-examination and passed with good results across the board.

Although Einstein still had some trouble with authority figures (the head of the ETH physics department, Heinrich Weber, once said to Einstein: 'You're a very clever boy, but you have one big fault: you will never allow yourself to be told anything'), his time at the ETH was thrilling academically. But now things were a lot less pleasant at home. With yet another failed business, his father had opted for straightforward employment – the pay was not wonderful, and money was very tight. Einstein dealt with this by cutting himself off as much as possible from his family life, concentrating wholeheartedly on the ETH. Before long he completed his independence by beginning to pursue a fellow student, Mileva Maric. Mileva wasn't his only female interest – the young Einstein had a string of girlfriends – but Mileva seemed to fascinate him, in part perhaps because she didn't return Einstein's interest until he had chased her for at least two years.

Although the science at the heart of his ETH course continued to fascinate Einstein, his rebelliousness surfaced again in a tendency to think it unnecessary to turn up for lectures that didn't interest him. The fact that Einstein managed to graduate was largely down to a friend, Marcel Grossman, who went to every lecture and made superb notes that Einstein increasingly relied on as the final examinations drew close. But Einstein did succeed and, inevitably, being Einstein, proceeded to try to take on academia his own way.

Rather than look for a graduate studentship, Einstein found a job teaching and tried to get a doctorate at the University of Zurich by writing scientific papers in his spare time. This early move into employment wasn't entirely inspired by an urge to be different. Having renounced German citizenship Einstein was stateless. If he wanted to become Swiss, he needed to have a full-time job. His attempts to get a post by writing to famous scientists proposing they took him on as an assistant came to nothing (though this was hardly surprising). The more mundane teaching job enabled him to take Swiss citizenship in 1901. But like Galileo before him, Einstein found that the life of a teacher got in the way of his ability to think. He found a very different role that would make it possible to earn a living and still have time for himself thanks to college friend and note-taker, Marcel Grossman.

The head of the Swiss Patent Office, Friedrich Haller, was a friend of Grossman's father. Einstein first contacted Haller at just the right time, when a new job was about to be advertised. Haller interviewed Einstein, and though it was obvious that his knowledge was more theoretical than practical, he took the young man on. The only slight irritation for Einstein was that after the interview Haller dropped the level of the post he was offering from Patent Officer (second class) to Patent Officer (third class) to reflect Einstein's lack of experience.

Light in the Patent Office

Sitting in his cramped office in the bustling little city of Bern, the 26-year-old Albert Einstein was enjoying life to the full for the first time since childhood. Bern was perfect – he had written to his fiancée Mileva before she joined him: 'It's delightful here in Bern. An ancient, exquisitely cozy city…'. Now he and Mileva had a family and Einstein had a steady job in the patent office. True, memories could still cause bleak moments. Mileva had given birth to their first child, a daughter, Lieserl, before they were married, when there was no hope of supporting the baby. What happened to their daughter has never been documented – she proved impossible to trace in later life – but she was probably brought up in Hungary by Mileva's family. Now, though, Einstein could comfort himself with the solid presence of his new baby son, Hans Albert.

The job was a godsend. Einstein sat back in his chair and took the next patent application from the tray. By the time he had read the first few lines it was obvious that the invention was totally flawed. He scrawled a note on the cover sheet and dropped the document onto a shaky pile on the floor. He had been amazed how easy the work had proved. He had never thought of himself as practical. While still at school, he had written in an essay that he imagined he would become a teacher in the theoretical sciences because he had a 'disposition for abstract and mathematical thought' and a 'lack of imagination and practical ability'.

Now, though, the ideas seemed to leap off the page, making it easy to filter out the failures. It gave him plenty of time for more productive thought. In the one year of 1905 he was to produce two world-shattering papers. The second is the one that made him famous, but the first, the one for which he won his Nobel Prize, was just as significant. And both involved light.

Today we have sophisticated photoelectric cells, used to power everything from calculators to space stations, but even back in the early 1900s it was known that shining a bright light on some metals would produce a small amount of electricity. The light hitting the metal was blasting some of the tiny electrons out of the metal's surface. Although light is insubstantial, it is partly electrical in nature, and this electrical component pushes against the electrically charged electron just as Faraday's wire was influenced by the electrical current in his crude motor.

This in itself was no surprise, but there was something strange about the way the electrons were produced. In 1902 the Hungarian physicist, Philipp Lenard had found a very odd result in his experiments. It didn't matter how bright or dim the light was, electrons knocked out by light of a certain colour had the same amount of energy. Move further down the spectrum of light and you would hit a colour where no electrons were produced at all. You would expect, if light were just a wave, that the more light you poured onto the metal, the more energy the photons would have. To make matters worse, there was also the ultraviolet catastrophe, the problem that had triggered Planck to come up with his theoretical packets of light.

By pretending that light's energy came in little packets, like sealed envelopes, Planck had predicted exactly how the photoelectric effect worked, but he didn't believe these packets existed. Einstein went one stage further, and accepted the unacceptable – that light actually was made up of these tiny packets. Somehow, Einstein thought, light managed to *be* a packet of energy, but still behave like a wave. It was left to American chemist Gilbert Lewis to give these packets the name we now use – photons – and another American, Robert Millikan, to prove that Einstein was right, but Einstein's theory explained away the problems. In this one step, thanks to the foundations Planck had laid, Einstein threw away one of the most basic accepted facts in science, that light was a perfectly normal wave, and left the way open for the whole of quantum theory, fundamentally changing our picture of the mechanics of reality.

Riding the sunbeam

Einstein's photoelectric effect paper was soon to be eclipsed, all because of a daydream. He and Mileva had been walking with Hans Albert between the manicured lawns of the city's park and now Einstein sat down on a grassy bank while his wife fussed over the baby. Einstein lay back and picked a piece of the grass, shredding it between his fingers. As he did so, he let the bright sunlight filter through his half-closed eyelids, enjoying the warmth of the Sun on his face. His lashes split the light into a hundred flickering beams. Einstein pictured the light itself, imagining it flowing through space like an incandescent river. He allowed himself to float with the river of light, riding on the sunbeam. It was pure relaxation.

Back in the office the next day, he tried to recapture the pleasure of that moment. What if he really *could* float on the sunbeam? What would the light look like? He stretched and stood for a moment, pacing the bounds of the office before returning to his high-backed wooden chair. He knew in principle what light was. The Scotsman, James Clerk Maxwell had described it in all its forms from radio waves to the penetrating X-rays with four compact equations. They showed light to be the interplay of electricity and magnetism, each generating the other. But this self-supporting miracle could not survive unless light moved at a particular, unique speed. Without it the electricity would not generate enough magnetism, the magnetism would not produce enough electrical current and the whole phenomenon would disappear. All this came out of Maxwell's equations.

Einstein got up to take another circuit round the crowded room, avoiding the piles of papers that littered the floor. There was something wrong. A contradiction between the theory and his daydream was glaring out at him, just as the flaws in the patents did. It was an irritation, an itch that he *had* to scratch. Maxwell's equations would only work if light were travelling at one particular speed – around 300,000 kilometres per second. But it wasn't enough to say that light moved at that speed. What was the speed being measured against?

When a train travelled at fifty miles an hour, that meant fifty miles an hour relative to the ground. But compared with another train running along at the same speed in the same direction, the first train would be stationary. For that matter, Einstein himself, sitting firmly in his chair, was obviously not going anywhere, yet at the same time, with everything else on the Earth, he was hurtling through space at thousands of miles an hour. The train's speed and Einstein's speed both depended on what they were measured against. So the same should hold true for light. In his daydream, when he was flying along with the sunbeam, it was entirely reasonable to say that the light was not moving.

If light didn't move at the right speed, its juggling act between electricity and magnetism would fall apart. The light would cease to exist. Either Maxwell and his elegant description of light were wrong, or Einstein's daydream was impossible. His gut feeling told him that Maxwell had to be right, and so it must be his journey on the sunbeam that was at fault. Against all common sense, light would not slow to a stop as he matched its speed. In fact, he never could match it, because however fast he moved towards it or away from it, light would still be rocketing past at 300,000 kilometres a second. Light's speed would not alter by a fraction whether he was stopped or moving. Unlike everything else in the natural world, light could only ever have the one speed, the unique speed of light.

Over the next few weeks, Einstein could think of little else. Once he knew that light's speed never altered, he was forced to re-think some of the most basic concepts in physics, dating back to Isaac Newton. When the strange, invariable speed of light was brought into the mathematics of movement, something had to give way. By simply working through the maths, Einstein realized that anything moving at near light speed would inhabit a bizarre world, where nothing behaved normally. His ideas stirred up a storm amongst scientists and public alike. The reluctant Einstein became a worldwide media phenomenon.

The implications of Einstein's daydream are mind-blowing. Reality stops working the usual way as you get close to the speed of light. Fixing light's speed has the effect of freeing up other factors that had until then seemed unchangeable. It's like trying to pin down an animal so a vet can give it an inoculation. Hold it by the legs and its head moves too much. Grab for the neck and the legs are suddenly kicking out. When Einstein pinned down light's speed, not only mass and size, but time itself, sprang free.

An object moving at near light speed shrinks and becomes hugely heavy. The passage of time for that object becomes detached from our own. As our clocks tick on, second by second, time for the object slows down, getting slower and slower until it stops entirely at the speed of light. If it were possible to move it even faster, to exceed light speed, the object would begin to move backwards in time. Since Einstein's first remarkable insight, all this has been proved by experiment.

The distortions of special relativity are most dramatic near the speed of light, but today's instruments can expose relativity's mysterious influence in the everyday world. Atomic clocks, slicing time into fragments of less than a billionth of a second, are small enough to fit into a suitcase. Take two of these hyper-accurate timepieces and synchronize them exactly. Fly one around the world, while the other stays firmly on the ground. Place the clock from the plane back alongside its earthbound equivalent. Compare the times now – and the clock that made the journey will have fallen behind, perhaps by 30 billionths of a second. While it was on the plane, time will have been running fractionally slower.

A frequent flier ages around one thousandth of a second less than a counterpart on the ground after 40 years of weekly Atlantic crossings. If the speed of light were a lot slower, the impact of special relativity on space and time would have been obvious all along. In a world where light only travelled at a quarter of a mile a second, that same frequent flier would have aged a year less than the colleague who never took to the air. It's the immense speed of light that stopped Newton's laws being questioned sooner.

Einstein had put light, and its speed, at the very centre of reality. With such amazing ideas he could not stay hidden away in the patent office for long. Einstein was soon receiving honorary degrees before he could even manage to get an academic post. By 1909 he was offered the new chair of theoretical physics at Zurich University and entered full-time academia. But this didn't imply slowing down. In fact, it was the same year, 1909, that he gave his first ever paper at a conference. Not only did he describe the way his earlier work made it seem that light managed to be a wave and a particle at the same time, he presented for the first time the equation that would be most firmly associated with him (and science) from then on, $E = mc^2$.

From here on Einstein began a carousel ride of academic positions – the University of Prague, his old college the ETH, the University of Berlin – continuing to output original ideas with remarkable consistency. Most scientists are like athletes, giving their best before they finish their 20s. Einstein was already 30 by the time he got his first academic post, and he was still to produce his masterwork, the general theory of relativity.

Distorting space

Where special relativity had ignored the effect of gravity, or in fact anything that was accelerating or decelerating, general relativity expanded the theory to deal with these more realistic cases. Just as special relativity had been sparked off by a daydream, general relativity began from a passing thought. Einstein later commented:

> I was sitting in a chair in the Patent Office at Bern when all of a sudden a thought occurred to me. If a person falls freely he will not feel his own weight. I was startled. The simple thought made a deep impression on me.

Again it was Einstein the mental observer that produced a new way of looking at the world. Imagining a person falling freely and feeling no gravity, for example in a falling lift, he made the leap of thinking that gravity and the effect of acceleration, of getting faster and faster as you do when you fall, are impossible to tell apart – are, in effect, the same thing. But this has a very strange consequence when applied to light.

Having decided that falling with a lift will have exactly the same effect as gravity, Einstein imagined shining a beam of light across the lift as it fell. From outside the lift, the light would clearly move in a straight line, but from inside the lift it would curve. In the fraction of time it took the light to cross the lift, the lift will have fallen a little, so the light will hit the far wall slightly further up than expected. As he had decided that the lift's fall and gravity produced exactly the same effects, Einstein deduced that light's path should be bent as it passes close to a heavy object and comes under its gravitational influence.

This seems quite removed from the strange deductions of special relativity, until you consider what is really being said. Einstein gives us a view of gravity that is just as much a relative one as the simple picture of two trains running alongside each other used in special relativity. The easy assumption is that light is 'pulled' out of its path by gravity. What general relativity says is a whole lot stranger. The light continues along in a straight line, but the space that the straight line runs through is twisted, pulled out of shape by the influence of gravity.

Imagine space as a thick sheet of rubber. A beam of light, a straight line of a different colour, is threaded through the rubber. Now put a heavy ball on the rubber. The rubber bends inward around the ball. Look at any segment of the light's path through the rubber and it is still moving in a straight line through 'rubber space' – only now it is bent around the ball. Technically the light travels in a straight line, but the space it passes through is pushed out of shape – it is warped. In 1915, at the Prussian Academy of Sciences in Berlin, Einstein published his general theory. Once again he was to become a darling of the media.

If Einstein's general theory of relativity were true, it should be possible to see light bending, but like special relativity, the effect is so small in normal circumstances that nothing shows up. However, near to the Sun, where the gravitational pull is very strong, the result should be obvious. A simple way to see whether or not Einstein is right is to check whether stars appear to have moved out of position towards the Sun when they are nearly in line with it in the sky. Unfortunately, it's impossible to see stars near the Sun, except under one very special condition – a total solar eclipse.

Total eclipse

In fact there could have been a proof of Einstein's ideas in place even before he produced his full paper. He had discussed the ideas as early as 1912 with a German scientist called Erwin Freundlich. It was getting the detailed maths right that took Einstein another three years. Freundlich led an expedition to the Crimea to monitor a total eclipse and to check for Einstein's warping in August 1914. His timing couldn't have been worse. The Russians, by now at war with Germany, caught Freundlich and assumed his complex telescopic equipment was intended for spying. He was held captive until the end of August, when an exchange of prisoners was arranged.

After the war, in 1919, English astronomer Sir Arthur Eddington led an expedition to Principe Island, off the African coast, to take measurements at a second eclipse. Once Eddington's team had set up their equipment in the mosquito-infested area they had been allocated and began to wait for the

day of the eclipse to arrive it looked as if the expedition would
be equally incapable of proving or disproving Einstein's theory.
Day after day was cloudy, hiding the Sun from sight. Even as
the Sun's disk began to be eaten away by the moon, clouds
still obscured everything but the Sun's direct rays. It was only
minutes before totality that the clouds parted and the team
managed to take a total of 16 photographs.

The triumph of success soon turned to misery. It seemed
that Einstein's theory was doomed to remain untested. As the
photographs were developed, plate after plate showed up blank.
For the first ten shots, thin cloud had still covered the Sun –
not enough to get in the way of the splendour of the eclipse,
but hiding the crucial stars. In the end, only two of the 16
plates were usable, but their evidence was enough. The stars
were shifted out of position – by just the amount predicted by
general relativity. The publicity from Eddington's expedition,
and the associated interest in relativity made sure that Einstein
would remain a public figure for the rest of his life. Such was
the hype now surrounding relativity that Einstein was invited
to appear for a season at the top British vaudeville theatre, the
London Palladium. The topic of relativity was surrounded by a
sense of mystery that here was a subject too complex for normal
people to understand. This challenge sparked public fascination
and debate worldwide.

Outside the success of the general theory, the period of
the First World War proved a nightmare for Einstein. He was
horrified by armed aggression and spent much time arguing the
pacifist cause – but to little effect. As the strain grew, he became
very ill. His crumbling marriage to Mileva headed inevitably
towards divorce. During the illness he lived in Berlin while
Mileva and the children stayed in Switzerland. A friend, Elsa

Löwenthal, nursed Einstein and in the process the two became very close. Unlike Mileva, Elsa had no interest in science – she seemed much more the ideal *hausfrau* that something inside Einstein pushed him to search for in a partner. In 1919, Elsa became his second wife.

Although general relativity is not as directly relevant to light's nature as special relativity, it was to provide a powerful new tool for optical science. Eddington's expedition has shown how the Sun, very slightly, bent the path of light. But compared to a whole galaxy, the Sun is a tiny, insignificant object. Imagine light streaming out from a very distant star. If it's far enough away, the chances of a single photon reaching the Earth are quite small. The star would be invisible. But then put a galaxy between us and the star. How would things change?

Before general relativity, the answer would be simple. Even if you had been able to see the star, it would now be hidden behind the vastly larger, brighter galaxy. But thanks to Einstein something different can happen. As the light from the star streams out around the galaxy, it is bent inwards. Just as light passing through a lens is bent inwards. Instead of the occasional random photon arriving at the Earth, the light from around the edges of the focusing galaxy is brought to a point. Where this point is not hidden by the galaxy itself, we can now see the star.

This remarkable effect, galactic lensing, has enabled astronomers to see further and further into space. It's not a universal solution. There has to be a galaxy in the right place, so the result is a series of tiny windows into the far reaches of the Universe. The distance that can be seen this way is truly amazing. The furthest objects detected are 13,000,000,000 light years away. The photons began crossing the Universe less than a billion years after the Big Bang. General relativity gives us not only a view to immense distances, but back to the farthest reaches of time.

By the early 1920s Einstein and his new wife were finding the increasing anti-semitism in Germany harder and harder to live with. In a schizophrenic way, Einstein the *great* scientist was feted – given a house near the river Havel by the Berlin authorities to honour his 50th birthday – while Einstein the *Jewish* scientist was ridiculed and slandered. In 1932 he left Germany, never to return. He settled at the Institute for Advanced Study at Princeton University in New York State, his academic home for the rest of his working life.

God does not play dice

When American physicist Robert Millikan proved by experiment that Einstein's groundbreaking 1905 paper proposing the existence of photons was correct, he was trying to prove Einstein wrong. Like many of his contemporaries, Millikan found the implications of the quantum theory that flowed from Einstein's original work too ridiculous to contemplate. It took him ten years of trying to break Einstein's argument before he realized he had in fact proved it. But Einstein himself was not comfortable with all that his theories suggested.

Einstein hated the uncertainty that underpins the quantum world. He thought it unnatural that reality should operate this way. 'The Old One', he said, 'does not play dice'. While he was happy that light did sometimes act as particles, he was determined to show that quantum theory was flawed. In one of his rare mistakes, Einstein not only failed, but also made the possibility of teleportation tantalizingly real.

It was 1935. Einstein had joined up with two younger
Europeans, Boris Podolsky and Nathan Rosen, also refugees
from the Nazis. At Princeton, the trio worked through the
mathematics of quantum theory from first principles, searching
out a failing that would prove it wrong. They came up with a
result so weird that it seemed like the proof they needed.

Quantum theory says that a light photon can exist in a
strange mixture of two possible states until it is measured – only
then does it decide which it's going to be. It's as if a child were
both a boy and a girl right up to the point it was born and it
was only at that moment that a coin was tossed and the 50:50
decision made. Einstein, Podolsky and Rosen's idea, called EPR
after their initials, takes this oddity even further. It's possible to
tangle two photons together in such a way that they are forced
to be opposites of each other when the measurement is taken.
If these photons were children, this tangling ensures that one
child will become a girl and the other a boy. It isn't decided
which is which, though. Either could be the boy until you look.
If 'your' photon happens to become the boy, the other instantly
becomes a girl.

Einstein imagined separating these tangled photons to a
great distance, then taking a peek at one. At this point, as if
a coin had been spun, it becomes one of the two possibilities. If it happened to be a 'boy', then instantly, however far
the distance that separates them, the other becomes a 'girl'.
Obviously impossible, he believed, as his own special relativity
showed that nothing, not even information, could travel faster
than light. Einstein was satisfied that he had found the flaw
that would eventually disprove quantum theory. But he was
wrong. Years after his death, Frenchman Alain Aspect made
the EPR phenomenon come true, demonstrating what has

been called the 'spooky link' between the two photons. And the implications, as we will see in Chapter 10, were even more extraordinary.

For the remaining 20 years of his life, Einstein put a huge amount of effort into trying to provide a theory that would bring together the working of all the forces of nature, so that electricity, magnetism, gravity and the atomic forces could all be explained in the same way. Like all his successors to date he failed, but it did not mean that he spent the time unproductively, contributing his thoughts to a wide range of projects. For the first time, these included military applications. Although he remained pacifist in principle, Einstein felt he had to support the Second World War because of the sheer evil of the Nazi threat. He even encouraged the US President Roosevelt to begin researching the atomic bomb, as he was concerned that the Germans would get a working bomb before anyone else could.

Einstein didn't have any direct involvement in the bomb project, though. The atom bomb might have depended in concept on his archetypal equation, $E = mc^2$, but there was nothing about bomb making in Einstein's theories. He was never interested in working at a practical level, and it's arguable that though he encouraged the President to ensure that the USA didn't fall behind, he still would have found it very difficult to contribute positively to a weapon of mass destruction.

In his last few years Einstein seemed truly to become the image of an absent-minded genius with which the media had branded him. On one occasion he had to ring up his office to make sure exactly where he lived. He had some difficulty persuading the office, which had strict instructions not to give out his address, that it really was Einstein they were speaking to. Early in the morning of 18 April 1955, in Princeton Hospital, Albert Einstein died. The Einstein legend is remarkable but, like Newton's, well deserved. Some Einstein stories could well be exaggerated or even untrue, but even as myths they give an accurate picture of the man – and nothing can dim the contributions he made to science.

Filtering the ripple

While the Victorians built many new technologies around light, the first half of the twentieth century was largely dedicated to the new understanding of light that flowed from Einstein's insight. One man though, the American Edwin Land, was to make a fortune out of a little-regarded property of light.

Back in the 1700s (see page 78), Erasmus Bartholin had noticed the strange way that Iceland spar crystals seemed to split light into two distinct varieties. The significance of this discovery had not been clear until Augustin Fresnel (see page 124) realized the implication of the way these clear crystals divided a beam of light. At the time it was yet to be fully proved that light was a wave, but Young and Fresnel were both convinced of it. And Young's idea that light was a 'lateral' wave like a ripple in a skipping rope was the key to Fresnel's understanding.

If light were such a wave, it would be possible to have light that rippled up and down, like waves on the sea, or side-to-side, like a rattlesnake's progress through the sand. Fresnel imagined that the Iceland spar contained an invisible grid of slots, some horizontal, some vertical. The side-to-side moving light would pass through the horizontal slots, but not the vertical. The up and down light would do the opposite. The result would be to split off the two types of light. If the crystal bent one set of rays more than the other, the result would be the two images that are actually produced.

As far as our eyes are concerned, there's no difference between light waves whichever direction they ripple in. The Sun's light is a jumble of rays with ripples oriented in every possible direction, leaving some to pass through each of the Iceland spar's grids. The other rays, rippling in directions between the two alternatives are pulled into line with the nearest direction. The orientation of a light ray's ripple is called its polarization and it was this phenomenon that made Land a millionaire many times over.

While he was student at Harvard in 1926, Land became fascinated with the nature of polarized light. Plenty was known about it by then. Certain crystals would select out light that was polarized in a particular direction. This was because the electrical part of the light wave fights against the electrical components of the atoms in the material. In a crystal with a strongly aligned structure, only light with a particular polarization that matches the crystal's alignment tends to pass through, just as if it were passing through one of Fresnel's invisible grids.

Reflecting in a mirror (or off the surface of a road) also cuts out some of the directions the light can ripple in, leaving it polarized, something that had been observed as early as 1808 by Frenchman Etienne Malus. At the time a colonel in Napoleon's engineering corps, Malus was idly playing with a piece of Iceland spar while staring out of his apartment windows at the Luxembourg Palace. He noticed that when the light reflected off the windowpane fell on the crystal, only one image was produced. Reflected light was naturally polarized, so the ripples of light that hit the Iceland spar were all in the same direction. There was nothing to be split off.

Land felt that the polarization effect had commercial value if only it could be strengthened and built into a more versatile material. Still only 18, he took leave of absence from Harvard and worked in a garage laboratory. The result of his experiments was Polaroid, a plastic sheet with tiny polarizing crystals embedded in it. By 1937, his garage laboratory had become the Polaroid Corporation. His hunch had paid off. Because reflection polarized light, the glare that irritated motorists or ruined photographs could be cut down dramatically by placing a piece of Polaroid material in the way. The material mostly let through light polarized in one direction (Figure 8.1); by rotating it to 90° from the polarization of the reflected light, Polaroid could remove glare from practically any surface.

Polarization was given a whole new lease of practical life much later in the century with the development of the liquid crystal. The liquid crystal rotates polarized light through 90°

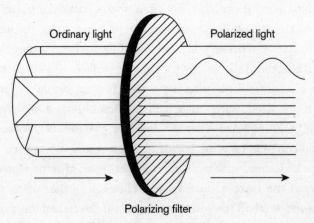

Ordinary light   Polarized light

Polarizing filter

Figure 8.1 A polarizing filter

– unless there is an electrical current across it. The crystal is sandwiched between two polarizing filters at 90° to each other. Normally that would mean that no light would get through – the result would be black. But the liquid crystal's 90° twist takes the light that passes through the first filter and gives it just the right twist to get through the second, giving a patch of light colour. When the power is on, the twist isn't applied, the light from the first filter is cut out by the second and the result is an area of black.

Polarizing materials continue to be important to this day, but in the second half of the twentieth century the technology of light was transformed from a specialist industry to an essential part of everyday life. Along the way, one man would make as great a contribution as Maxwell and Einstein to light's complex history.

Chapter Nine

QED

The world we live in is but thickened light.
RALPH WALDO EMERSON

Ask a person in the street to name the two greatest physicists of the twentieth century and they will almost inevitably come up with Einstein. The second name, though, might prove harder to pin down. Ask a physicist to come up with the top two and there will be no hesitation. Or at least, if there is any hesitation, it will be over which name to put in first place. The name that ranks alongside Einstein will be that of Richard Feynman.

The ultimate showman

That Feynman does not have the mystique of Einstein is not because the man was a shy, retiring academic. He was anything but. The word that is most often applied to Feynman's style is 'showman'. Feynman was not only a genius for whom untangling the mechanisms of the Universe was a joy, he also had a remarkable ability to captivate audiences and share his excitement.

This showmanship is so much at the heart of the way Feynman operated that it permeates his thought. Before plunging into the transformation he made to light's lifecycle, it's worth spending a few moments on his most famous example of show-manship, which gained him a worldwide audience towards the end of his life. Such stories are an essential part of understanding Feynman. He was himself a superb storyteller, considering care-fully crafted anecdotes a fundamental part of his craft.

On 28 January 1986, around midday Eastern Standard Time, the space shuttle *Challenger* took off from Cape Canaveral. Around a minute later, the rocket booster exploded into a ball of fire, killing all the astronauts. The TV pictures echoed around the world, establishing one of the most memo-rable media images of the twentieth century. Feynman was part of the commission set up to investigate the accident. Unlike the other commissioners he had no connection with the space industry; he was there because the Acting Head of NASA, William Graham, who drew up the list of names, had attended Feynman's lectures in the 1960s, and like everyone who had heard Feynman speak, was a fan.

Feynman hadn't wanted to take on the job. He had already contracted cancer and was suffering from heart problems. But his wife, Gweneth, persuaded him that it would take someone with his unconventional approach to get past the bureaucratic smokescreen. Feynman's style was certainly very different from that of the civil servants and military experts who dominated the commission. He was frustrated by the slowness with which they operated and the lengthy, time-wasting fact-finding visits. Like Einstein he had no love of the paraphernalia of authority. Spurred on by an engineer to consider the effect that the freez-ing temperature at the time of the launch might have on the huge O-shaped rubber rings that sealed the joints in the rocket motors, he was not willing to let the process of investigation drag on for months.

At a televised session of the commission that was only intended to revisit old ground for the camera's benefit, Feynman sprang a surprise on the bureaucrats. He captured a model of the joint, complete with O-ring, that was being passed around the table. Once the cameras were on him, Feynman was ready to act. He slipped a pair of pliers and a screwdriver out of his pocket and used them to extract a piece of the rubber. This he compressed with a small clamp, much as the rubber would have been squeezed by the great shell of the motor. Then he plunged it into the iced water that had been provided for the committee members to drink. When the rubber was removed and undamped, instead of springing back into shape it took several seconds to recover – Feynman had demonstrated graphically that the O-rings would not maintain the flexible seal that was essential for safety at around freezing point.

Seeing beyond the labels

This ability to get across the scientific message without putting off the audience would have made Feynman a great communicator even if he hadn't been much of a theoretician, but there was much more to him. Something Feynman discovered as a child and never lost was a fascination with the way things worked. He always attributed this to his father. When Melville and Lucille Feynman had a son on 11 May 1918, they couldn't have predicted his future, but Melville was determined that his boy would see beyond the labels that so often are used to demark and limit possibility. He encouraged Richard to think about the nature of things, not the labels we apply to them.

Feynman's lack of enthusiasm for formality and boundaries is legendary. This not only led him to ignore labels and look for the reality beneath, but also gave him the rare ability to have the breadth of interest of a da Vinci or a Newton in a world of tighter and tighter specialization. Most physicists restrict themselves to a very narrow slice of the scientific world. This way they can claim expertise and cling onto a measure of status as *the* source on a particular topic. It also means that they can apply the lengthy thinking and development necessary to open up new avenues of science.

Feynman had little interest in status (when awarded the Nobel prize he seriously considered turning it down, and he never accepted an honorary degree). He genuinely didn't care about being acknowledged as the first to pin down a theory, which meant he was often slow to publish his findings; it was enough that *he* knew. With this lack of status-consciousness came an unusually concentrated way of working – he would often develop his theories in a very short, intense session – and a wonderful breadth of interest. Any and every topic pertaining to physics interested him. He would quite happily skip from challenge to challenge as long as there was a puzzle to solve.

In fact he didn't even limit himself to physics. He regularly dabbled in biology, a habit that dated back to his days in graduate school, when to broaden his experience he had attended undergraduate biology lectures. A story he tells of his experience in his classic collection of anecdotes, *Surely You're Joking, Mr Feynman*, typifies the way Feynman was never happy building on other people's work, but instead liked to start everything from basics, an approach that would make possible his unique contribution to physics. One of the class exercises was to comment on a paper in which the electrical impulses in the nerves of a cat had been measured. Feynman wanted to put the various parts of the cat mentioned into context, so went to the library and asked for a map of a cat, much to the librarian's confusion.

When Feynman then reported on the paper, he began by explaining just what was what inside the cat. His fellow students pointed out they knew the names of the cat's muscles already. 'Oh', said Feynman, 'you do? Then no wonder I can catch up with you so fast after you've had four years of biology'. Tact wasn't one of his strong points. As far as he was concerned they had wasted their time memorizing stuff that could be looked up in fifteen minutes. Later, in his famous 'red books' document-ing a series of undergraduate physics lectures he gave in the early 1960s, he included a summary of the whole of classical physics that took up less than half the (admittedly spacious) page. Practically everything else was just 'stuff that could be looked up in fifteen minutes' or could be deduced from those nine equations.

Discovering physics

After a successful time at high school, Feynman applied to two universities – Columbia and the Massachusetts Institute of Technology (MIT). Columbia wouldn't take him as it had already filled its quota of Jews (this was 1935, when racism was still endemic in the USA). MIT did want him – and Feynman began a very successful four years as an undergraduate. He started out in mathematics – Feynman's experience from school was that maths came as easy as walking or eating – but before long he realized that serious mathematicians almost prided themselves on the lack of application of their work. Wanting something more practical, he switched to engineering, but found this lacking in challenge. Finally Feynman settled on the long-term love of his life, physics.

For many undergraduates, college days involve getting through the minimum amount of work needed to complete the course and otherwise having a good time. Although Feynman wasn't a social outcast, physics dominated his life. He was soon taking courses aimed at more advanced students to expand his experience, and in his final year had two papers published in the respected journal *Physical Review*, a singular honour for an undergraduate. There was no doubt in his mind what he wanted to do after graduation. The only question was which graduate school to apply to, and with Feynman's reputation going before him, he had plenty of choice. He opted to stay where he was.

Feynman was a product of his culture, of late 1930s America, a country that was at that time both parochial and isolationist. His views on many matters of importance to the world would always remain childlike. He was uncomfortable with the concept of literature, dismissive of philosophy and religion, and could never understand the attraction of music, except drumming, which he loved. This cultural immaturity made staying put at MIT the obvious option – it was the place he knew, so it had to be right for him. Luckily the head of physics at MIT, John Slater, had a less narrow view and insisted that Feynman moved on to see more of the world. He was to choose Princeton.

Princeton accepted Feynman despite truly abysmal results in his graduate admissions assessments in English and history. There was also some concern about his being Jewish. Princeton suffered from excessive airs and graces at the time. Modelled on the colleges of Oxford and Cambridge; it was a very formal Ivy League institution, and though the school had no Jewish quota, its admissions board could be decidedly sniffy about a candidate's origins. It was only assurances from Feynman's lecturers and supervisors at MIT that convinced the Princeton authorities that he was a catch that shouldn't be missed – and that he neither looked nor acted like a Jew.

The more formal aspects of Princeton life just weren't Feynman. But the starched behaviour expected in the graduate college didn't wash over into the physics department, which was of top quality – boosted by the close proximity of Einstein, based at Princeton's Institute for Advanced Studies. After looking into a number of different physics problems, Feynman settled down on a PhD thesis in the field of quantum mechanics. This was to be a major step along the road to his Nobel Prize winning work that would transform the position of light in the substance of reality. Feynman wanted to find a simpler approach to the description of quantum theory, which up until then had required complex mathematics to follow what was happening. The trouble was, he couldn't get a toehold. He didn't know where to start.

World lines

It wasn't until Spring 1941 that he accidentally came across a paper by British physicist Paul Dirac describing how a very small portion of what Feynman wanted to do could be achieved. In one of his remarkable leaps of intuition, Feynman saw how Dirac's work could be generalized. The result, as often was the case with Feynman, was a very visual approach. This involved considering the possible 'world lines' of a particle. These are diagrams where the particle's position in space is plotted against time (Figure 9.1). If time goes up the page and space across, a

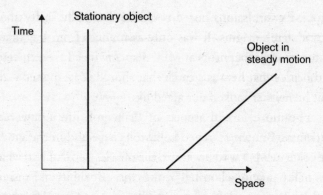

Figure 9.1 World lines of a stationary and steadily moving particle

particle that is not moving at all will be described by a straight line up the page. One that is moving at a steady speed will be shown as a diagonal straight line. The world line is a map of the history of a particle's existence.

Feynman saw that he could fully describe a particle's behaviour by drawing every possible world line linking its starting and finishing position and pulling them together with the probability of such a line occurring. This sounds an impossible task, but ever since Newton and Leibniz had come up with calculus it had been routinely possible to combine an infinite number of components, provided the values got smaller and smaller. For the moment, this discovery did not have any practical use, but it was an excellent basis for his thesis. Work on his PhD progressed smoothly until the Second World War suddenly got a whole lot closer. In the summer of 1941, before the Japanese attack on Pearl Harbor in December, Feynman had already got involved in war work, taking a summer job working on a mechanical computer to help make bomb delivery more accurate. With the full-scale commitment of the USA to the war, his life was to undergo a much bigger change.

The same month as the devastating attack on the US Pacific fleet, Feynman was asked to join a new, top-secret team. Since Einstein's warning letter to the President, there had been concerns that the Germans were developing a new kind of explosive device, one that made use of the unstable nature of the uranium isotope U-235 to generate a huge blast – the atomic bomb. The USA and its new allies felt that the only safe move was to develop their own bomb, and to develop it quickly. Feynman turned down the offer. He disliked the unthinking procedure of the military, he had his PhD thesis to finish, and he was no great enthusiast for mass destruction. But as the afternoon ticked away he could not put the thought of the bomb out of his mind. Of what the Germans could do if they developed it first; of what the Germans were already doing to so many fellow Jews. He gave up the struggle and within minutes was established in his new role.

Towards Trinity

The team Feynman found himself in was not actually building the bomb, but looking at methods of getting hold of U-235 in the first place. In its natural state, uranium mostly consists of the stable U-238 variety, with a very, very small percentage of its radioactive cousin scattered amongst it. Separating out the U-235 was not a trivial task. This was important work, but did not stretch Feynman. As things were going slowly, he managed to get a few weeks off to finish his PhD thesis and submit it, receiving his doctorate in June 1942. Despite his lack of enthusiasm for status, Feynman had a pressing reason for wanting his PhD. His grant as a graduate student only continued while he was unmarried. With this restriction lifted, he and his long-time fiancée Arline were able to set a wedding date for the following month.

The haste was more than just youthful enthusiasm. Arline had been seriously ill for some time, and after a misdiagnosis of Hodgkin's disease, she was discovered to have tuberculosis of the lymph glands. The doctors gave her only a few years to live. Feynman's family tried to persuade him not to go through with the marriage. They were all fond of Arline, but felt Feynman was tying himself down unnecessarily, even putting himself at risk of catching TB. As it was, his physical relationship with his new wife had to be strictly restricted to avoid infection, but Feynman would not give up on Arline. He arranged for his new bride to be moved to a hospital near Princeton (she would spend all their marriage in a hospital bed), and even rigged up his own mini-ambulance to take her out of the hospital for the wedding.

By the end of 1942 it was obvious that the attempts of Feynman's team to separate U-235 were going to be made redundant by the method of a competing group at the University of California. Feynman, along with the whole team, was asked to move down to Los Alamos in New Mexico to join the Manhattan Project – the immense effort focused on building the bomb. He agreed, with the provision that Arline be found a hospital bed nearby – in the event she was moved to Albuquerque, about 60 miles away as the crow flies.

Feynman's contributions to the Manhattan Project were surprisingly varied. He had a determination to succeed, whatever action that implied. He was just as happy mending calculators as working on theory, a skill that got him put in charge of the theoretical computations group, using sophisti-cated electro-mechanical IBM calculating machines to process the miles of numbers needed to detail the construction of the bombs. Feynman was given this important position despite having just graduated because of his quick thinking and lack of regard for authority. When Hans Bethe, the head of the theory division, wanted to throw some ideas around, it happened to be Feynman he chose to bounce them off. Rather than be deferential, Feynman told Bethe exactly what he thought – when he agreed and when he didn't. The older man appreciated Feynman's strength of character and rewarded his insubordination by giving him a team leader position.

As work on the bomb continued, Arline's health deteriorated. The couple kept up a steady flow of letters and Feynman made the trip to Albuquerque most weekends, but by the spring of 1945 Arline was getting very weak. She died in June. One month later, Feynman was summoned back to Los Alamos by a cryptic message from Hans Bethe, saying that 'the baby is expected'. On 16 July 1945 the atomic bomb test, codenamed Trinity, took place. A plutonium-based bomb was rigged up at the top of a 110-foot tower in the desert near Alamogordo, 200 miles south of Los Alamos. Feynman was present to see the first ever nuclear explosion. Less than a month later the uranium-based bomb was dropped on Hiroshima and then a plutonium bomb on Nagasaki, bringing the war and the Manhattan Project to an end.

Seeing the world differently

Strangely, though, it's not for any of his work as a part of the Manhattan Project that Feynman is best remembered, but rather for his extra-mural activities. From his first days at Los Alamos he set out to make life difficult for the security team.

It wasn't that Feynman didn't see the need for security in a war, but he was appalled by the classic bureaucratic mentality that considers action more important than substance. Most of the secret papers detailing how to extract uranium and build a bomb were kept in filing cabinets, locked with ordinary, corner store padlocks. It wasn't Feynman's style to issue a critical memo. Instead he picked the padlocks, opening the filing cabinets to demonstrate how vulnerable they were. He even found that by tilting a cabinet over he could extract the contents of the bottom drawer without unlocking it. If he needed a report from an empty office, Feynman would pick the lock, take out the document and relock the cabinet, making sure that his unwary benefactor found out just what had happened.

Eventually the irritation of Feynman's constant pressure spurred the security office into coming up with an answer. It's easy to imagine them thinking 'this'll fix him' as they installed a new batch of high-security cabinets fitted with the Mosler Safe Company's combination locks. This time Feynman couldn't come up with an instant solution. The locks required three numbers to be entered in sequence. There was nothing to pick, no way to hear the tumblers fall. But by experimenting with the locks, Feynman soon discovered that the combinations were less fearsome than they seemed. To enter the number 65, anything between 63 and 67 would do, reducing the possibilities on each turn of the dial to 20. This, reckoned Feynman, would make it possible to get the cabinet open with around four hours' mechanical effort. But that was too much like hard work.

It was a chance discovery that got Feynman past the locks. When a cabinet was open, he found that it was possible to read off the last two numbers of the combination. The lock bolt would twitch as he turned the dial past the crucial positions. Before long he could pin down the numbers by touch with his back to the cabinet, just glancing at the dial to note the result. At every possible opportunity he noted the second and third numbers of open cabinets. Then, when he needed to get past the lock, he only had to try the 20 possible first numbers. The Mosler Safe Company was beaten. Once he had a puzzle to solve, he would worry it like dog, coming at it from different angles, never letting up until it had given way. It was the same part of Feynman's character that made him such a great physicist. It was lucky for the US and its allies that there weren't any spies around with Feynman's combination of genius and persistence.

At the close of the project, Feynman had pretty much his choice of universities. He decided to accompany his mentor at Los Alamos, Hans Bethe, back to Cornell University at Ithaca, New York, like Princeton one of the distinguished Ivy League schools. For the first year or so, he was very depressed. The implications of his work on the atomic bomb had not escaped him, he had lost Arline and in 1946 his father also died. But Feynman could not resist the siren song of physics for long. The opportunity of venturing further into the reality that lay beneath light and matter was too exciting.

At the heart of much of Feynman's original thinking is a simple but powerful concept – a combination that scientists often label 'elegant' – the same concept that Fermat had used when uniting the mechanisms of reflection and refraction, the principle of least action. This is the principle that says that creation is lazy, that a physical phenomenon will take the course that requires least effort or least time. Feynman was 16 when his physics teacher, Abram Bader, introduced him to this principle. The same principle lay behind his PhD thesis on quantum mechanics, and the theory with which his name will always be linked – Quantum Electrodynamics, QED for short.

The strange theory

Feynman himself was later to describe QED as 'the strange theory of light and matter'. In essence, QED describes how light and electrons interact. This isn't just of passing interest – it's fundamental to an understanding of light. Light doesn't come out of nowhere. A photon begins life, travels – potentially for billions of years – and then is destroyed. Each end of its existence, birth and death, involves an interaction with matter. It's the electrical charge of the electron that creates the photon in the first place, kicking off the elegant dance between electrical and magnetic components that Maxwell first described. And it's another electron's electrical field that wipes it out at the end. QED completes the life cycle of light. And, as Feynman was to discover, light is fundamental to the continued existence of matter.

Feynman was not alone in developing QED. Although his PhD thesis provided the tool that would be necessary to make any practical progress, another young professor, Julian Schwinger, based at Harvard, was to make many of the practical steps. There was further input from Bethe, and Schwinger's work was independently duplicated by Japanese scientist Sin'Itiro Tomonaga. Even so, Feynman's ability to put across his ideas in a clear pictorial form made all the difference. This wasn't immediately obvious – in fact, Feynman's pictorial approach was alien to many physicists, but a young Englishman, Freeman Dyson, who was studying under Bethe, managed to pull together Tomonaga, Schwinger and Feynman's approaches to QED in a single review that finally enabled the physics world to appreciate just what a contribution Feynman's technique made.

QED is remarkable in many respects. With the exception of gravity and the workings of the nucleus of atoms, QED describes how everything works, full stop. From it, with sufficient patience, it is possible to derive pretty well all of physics and chemistry that doesn't fall into those areas of exception. What's more, unlike many physical theories, QED has a stunningly close match to reality – and even the error that does exist is understood: it is merely the exact calculation that is beyond practicality.

QED takes the workings of light down to individual photons, which are small enough to be subject to quantum theory. There's no doubt at all that photons exist. As Feynman pointed out in a series of popular lectures on QED, it is now simple enough to produce and measure the impact of a single photon. Feynman left no room for doubt in his book on QED:

> I want to emphasize that light comes in this form – particles. It is very important to know that light behaves like particles, especially for those of you who have gone to school, where you were probably told about light behaving like waves. I'm telling you the way it *does* behave – like particles.

So in principle, according to Feynman, Newton was right in his query that suggested light was a series of corpuscles. But being quantum particles, there's some pretty strange behaviour going on.

Quantum behaviour

Stand inside a lit room and look at a window at night. You will see a mixture of images – partly what's outside, partly the reflection of what's inside. The glass is letting some of the light pass through, while some of it is bouncing back. So let's imagine a beam of light particles hitting the glass. We count the photons that are reflected and the photons that pass on through. In any particular case there will be a certain number going each way. There's a probability that the photons will be reflected or pass through. But how does each photon decide which to do?

Newton had real problems with this, and it's one of the main reasons that the wave theory succeeded in knocking Newton's out of the way. He and others after him played around with explanations for the way that seemingly arbitrarily some light particles would reflect and some would pass through, but no argument could be found that would stand up to experiment. We know that a certain percentage of photons will bounce off, but not how an individual photon will behave. It's as much a game of chance as roulette (in fact more so, as roulette wheels are not truly random). That nature should behave in this arbitrary way made Newton uncomfortable. He wanted to find an absolute rule behind the mechanism, but arbitrary it has remained.

However, if things seem odd when looking at a photon's choice of whether or not to bounce off, they get absolutely crazy when you consider what happens at both sides of the glass. Let's stay with that picture of looking at a window at night. Some of the photons from your side of the glass bounce back to give a reflection. Others carry on into the glass. Soon they have reached the other side. Again there's a choice. They might carry on straight into the cold night air, or they could bounce back into the glass, reflected from the far side. We know light can bounce off the edge of glass where it meets the air – this is how fibre optics work.

So far, this doesn't seem too remarkable. But something really weird is happening when you get down to counting the individual photons. The number of photons that get reflected depends on how thick the glass is. That's not entirely surprising, as you could imagine that somehow the glass's thickness could change how many photons bounce off the back of the glass where it meets the outside air. But what's strange, very strange, is that the number of photons bouncing off the *front* edge of the glass is also changed by the thickness. It's as if the photon knows at the point it bounces off the front just how far it would have travelled through the glass, if only it had done so. Spooky indeed.

The way light reflects off two surfaces is explained with no problem by the wave theory – it's our old friend interference, taking place between the waves bouncing off the far side of the glass and the nearby ones. But how can this work if light is made up of individual photons? Yet the fact is, it does, even if the experiment is conducted a photon at a time.

Remember, for that matter, Thomas Young's pair of slits, producing a pattern of dark and light bars. Again, it's easy to explain this if light is a wave, because the two waves can interfere with each other, adding together or cancelling out to produce the bright and dark sections. But how can individual photons work like this? And they do – sent through one at a time they still build up an interference pattern, as if each photon somehow manages to go through both slits and interfere with itself.

The icing on the cake of the unreasonableness of quantum behaviour is that the photon refuses to be caught out. If you put special detectors in the slits that count how many photons go through each, the interference pattern disappears entirely. Once you prove which slit an individual photon has gone through there's no more interference. Leave it a probabilistic mystery and the pattern reappears.

Feynman has good news and bad news. Quantum electrodynamics, QED, is a theory that accurately predicts what actually happens – but it doesn't make what's happening seem any more reasonable or logical. It's just a fact of life that our brains deal with the normal world, not the weird quantum level, and they simply can't function any other way.

Arrows of time

Feynman's secret weapon in describing what was happening was his visual mind. Rather than work with vast quantities of featureless equations, he liked diagrams, he *thought* in diagrams. Feynman developed a set of diagrams covered in little arrows, where the size of the arrow indicated the chance of a particular event happening and the direction of the arrow indicated the point in time, making the arrows rotate with time like the second hand of a clock. By combining all the arrows for the possible ways a photon could behave he could accurately predict its behaviour.

The more he thought about these little arrows of probability, the more Feynman could see that all the complexities of the behaviour of light – reflection, refraction, interference, diffraction – that seemed to require waves to explain them could be explained purely in terms of photons, provided you realized just how strangely photons behaved. This doesn't say that light hasn't got wave-like properties – and most physicists today still find it convenient to regard it as a hybrid of wave and particle – but Feynman and his colleagues produced a theory that could explain everything light does without ever needing a wave.

One of the reasons that the wave remains a popular description of light is that thinking the QED way means coping with the weird quantum view of the world. Take that simplest action of light, reflection. We're used to the way 'real world' things bounce. A ball, for instance. When it hits the floor at an angle, it bounces off in the opposite direction at the same angle. The same thing happens to a wave on water. So it seemed reasonable to assume that light behaved the same way – and so it appears to do. But QED makes *appears* the key word. Those irritatingly probabilistic photons are no respecters of the expectations of the ordinary world.

In fact, when a photon hits a mirror at a particular angle it could reflect off at any old angle. Imagine a beam of light, hitting a mirror and bouncing up to your eye. QED says it doesn't have to travel to the middle of the mirror and reflect to your eye at the same angle. It could hit anywhere along, then bounce up at a totally different angle to reach the eye. But when you add up all of Feynman's little probability arrows for the different routes, most of them cancel each other out. The final outcome is to travel along the path that takes the least time – reflection at equal angles.

But just because all those other probabilities are cancelling each other out doesn't mean they don't exist. And you can prove this. If you chop off most of the mirror, leaving only a piece to one side, you obviously won't get a reflection. But put a series of thin dark strips on it, only leaving available those paths whose probability arrows are pointing in the same direction and it begins to reflect, even though the light is now heading off in a totally inappropriate direction for reflection as we understand it.

You can actually see this happening without bothering to fiddle around with mirrors and fine lines. Remember Feynman's little arrows – they rotated with time. But they rotate at different speeds depending on the frequency of the light. In visible light terms, that's the colour of the light. So different colours will be reflected off-angle to a different degree by these fine lines. Shine white light on a special mirror with fine lines engraved on it and you should see rainbows. Practically everyone has a mirror like this – a CD or DVD. Turn it over to see the shiny playing side and tilt it against the light. The rainbow patterns you see are due to the rows of pits in the surface cutting out the little arrows for one direction, leaving light reflecting at a crazy angle into your eye. QED in the home.

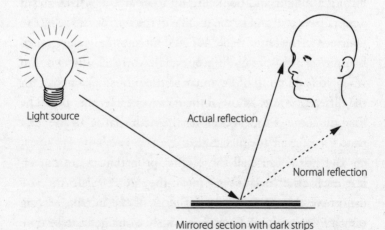

Light source   Actual reflection

Normal reflection

Mirrored section with dark strips

Figure 9.2 Reflection at an unexpected angle predicted by QED

This same approach can be used to work through all the behaviour of optics. For example, the way a lens changes in thickness from edges to centre means that all the probability arrows rotate just the right amount to arrive at a particular point, the focus, all pointing in the right direction. But QED does more than explain what's happening with light using these arrows. Because light isn't all that's involved. In all of light's behaviour, except for flying along in a straight line, it's interacting with matter. To be specific, it's thanks to a charged particle like an electron that a photon of light begins or ends its life.

Dancing with electrons

With electrons and photons in the picture, Feynman's diagrams became a little more complicated. To understand what was happening, he returned to world lines, those graphs with time running up the page and space across. By drawing diagrams for each possible interaction of electrons and photons and combining this with his probability work, it was possible to predict successfully practically everything that happens (except anything involving gravity and the centre – the nucleus – of an atom).

By bringing in the electron, we can understand the strange way that QED says light reflects from a mirror. It's only crazy if we think of reflected light bouncing off a mirror, just as a ball bounces off the floor. But thanks to QED, we know that each photon is absorbed by an electron in the mirror, and then a new photon is emitted in a different direction (different because the probability arrow rotates during the brief time the process takes). Photons don't have to bounce normally, because they aren't bouncing at all.

With this picture, the mystery of reflection from both sides of a piece of glass, that means we both see through it and see our reflection, goes away. When Newton was tearing his hair out trying to explain how light particles bouncing off both faces could be affected by the thickness of the glass, he had missed what was actually happening. In practice, electrons all the way through the glass catch hold of photons and then push out new ones. The combined probabilities produce the part-reflection. As this process happens throughout the thickness of the glass, it's not surprising that the thickness changes the way in which reflection occurs.

A similar process of capture and release is happening in our old friend, the atmospheric scattering that produces the blue of the sky. A photon from the Sun is taken in by an electron in an air molecule. A moment later the electron emits a new photon, but in a different direction – it is scattered. And this ability to absorb and re-emit photons allows physicists to explain another mystery – why we don't all collapse into nothing.

All matter, each of us, is made up of atoms. Each atom consists of a nucleus, a compact, relatively heavy part that has a positive electrical charge, and one or more much lighter electrons, which are negatively charged. Positive and negatives attract each other, so it seems reasonable that the electrons would plunge into the nucleus and get wiped out. Thankfully this doesn't happen, and it is prevented by light. The nucleus and the electrons are constantly exchanging a flow of photons, providing just enough interaction to keep the electrons in their place. This means that each atom, each physical object, each one of us contains an absolute fireball of light, busily ensuring that matter stays intact. It's not visible – it can never get out of the tight world of the atom – but it's there inside us: we are truly creatures of light.

QED underlies almost of physics and chemistry at a fundamental level. It explains the behaviour of light without ever resorting to waves, and shows how a network of light keeps all of us in existence. Feynman's diagrams are as much a part of modern understanding of QED as they were in the 1940s. His contribution to our understanding of light and matter is unrivalled in history.

Settling down

By 1950 the coldness of New York State and the low esteem in which science was held at Cornell were starting to get Feynman down. He toyed with living in South America, but ended up moving to the California Institute of Technology (Caltech) at Pasadena. In the years since Arline's death, Feynman had gained something of a reputation as a playboy. Now he felt the need to settle down. After a whirlwind romance he married Mary Lou Bell. Feynman met Mary Lou when she was an art history student at Cornell. Their attraction was one of opposites. Feynman's life with Mary Lou was anything but boring – in fact it seems to have been one long argument.

Mary Lou had very clear ideas of what a professor should be like, and tried to force the nonconformist Feynman into her stereotyped image of a respectable savant. At the same time, she thought physicists dull, and tried to encourage Feynman to keep his social life separate from work. She wanted him to give up the parties he delighted in and move in more refined, artistic circles. This behaviour was typified by her treatment of the great physicist Niels Bohr, who had hoped to meet Feynman while briefly visiting Caltech. Mary Lou informed Feynman when it was too late to do anything about it that he had missed out on dinner with 'some old bore'.

During his marriage to Mary Lou, Feynman aptly moved his main area of study into a very cold topic – superfluidity. When helium gas is cooled to the region of –271 Celsius, only two degrees above the absolute minimum temperature, it begins to behave very strangely. Its electrical resistance disappears, making it a perfect conductor of electricity. It develops odd habits, climbing up tubes as if trying to escape in defiance of gravity. Feynman generated a flurry of papers, explaining what was happening to the atoms in this unlikely substance, bringing to play once more his remarkable diagrams.

The Feynmans' marriage lasted until 1956, around four years in all. Mary Lou had singularly failed in the role of a replacement for Arline that Feynman had planned for her. It wasn't that Arline had been weak. She had frequently stood up to Feynman, but her stance had always been to encourage his originality. When Feynman complained about the way other people reacted to his ideas, she delighted in telling him 'What do you care what other people think?'. Where Mary Lou's arguments with Feynman had been destructive, Arline had only disagreed in a positive way, building up his confidence. It was an essential approach to life for anyone who was to fill some of the gap Arline had left, and Feynman found it in a young Englishwoman he met in Switzerland in 1958.

Gweneth Howarth, then only 24 to Feynman's 40, had sufficient independence of spirit and sense of fun to match Feynman's own. She had quit her job and set off practically unfunded on a round-the-world trip, showing remarkable daring for a young woman of the time. But she had run out of cash quicker than expected. Now she was working at subsistence levels as an *au pair* in Geneva. Feynman got on very well with her and made her an easily misunderstood offer. He suggested she came out to California as his housekeeper, where he would give her a significantly higher rate of pay, enabling her to carry on her travels at a later date if she wanted to.

It wasn't an easy decision for Gweneth, but she had already proved she didn't care much about the opinions of onlookers, who in the late 1950s were bound to disapprove of her actions. She moved to Feynman's house in Altadena, California in June of the next year. For a while they lived separate lives. Gweneth had boyfriends, Feynman his usual string of dates. Just occasionally he might go out with Gweneth, and they'd have a good time, but it was still a surprise for her when he proposed. They married in 1960, a marriage that was to last the rest of Feynman's life.

From shortly before meeting Gweneth, Feynman had been gripped by a new problem, that of the weak nuclear interaction, where the nucleus of an atom fires off an electron, Rutherford's beta particle. Applying his diagrams once more, Feynman managed to advance the understanding of this fundamental interaction to the extent that he well deserved a Nobel prize for his work. In fact (not that Feynman cared) it was for his 1940s work on QED that he received the prize in 1965.

Feynman worked constantly until his death in 1988 with little of the tailing off that occurs in most scientists. He proved himself an unparalleled teacher, giving lectures to packed audiences in his distinctive drawl – imagine Tony Curtis giving a physics lecture. After his work on the weak nuclear force he went on to contribute to our knowledge of the structure of apparently fundamental particles like electrons and protons, and made major steps in our understanding of gravity, once more of Nobel Prize stature, before cancer ended his life at the age of 70. At the same time he achieved his long-term aim of family life, into which he threw himself with just as much energy as he did with physics. Thanks to Feynman a whole generation of physicists have grown up with a special affection for the subject – and those of us without the technical knowledge to understand the details can still be amazed at the fundamental role of light in all matter that Feynman uncovered and described.

With Feynman's extension of the quantum revelations of Planck and Einstein, the building blocks were in place to construct the most remarkable technology ever envisaged.

A new order

At the heart of much of this technology was a single invention – the laser. The laser is a child of mixed parentage. Its initial form, working on light in the microwave spectrum, was an accidental discovery by Russian scientists Nikolai Basov and Alexander Prochorov. The Russian team was investigating the behaviour of the pungent gas ammonia, early in 1954. Some 37 years earlier, Einstein had predicted that it would be possible to set off a chain reaction producing light, which he described as stimulated emission.

According to Einstein's theory, an electron in an atom can be pushed into a high energy state when it is hit by a photon, leaving it like a bucket of water sitting over an open door. Another photon, hitting that electron, would not only be re-emitted itself, but would trigger the electron to release the stored up energy as a second photon – as if the bucket was knocked off the door by the stream of water from a hose, resulting in a doubled downpour.

Basov and Prochorov found that light of the right frequency, in the non-visible microwave region, triggered the release of more photons from ammonia. Generated in a sealed chamber, those photons could themselves stimulate further photons, a pyramid selling approach to producing light. Unlike a conventional source of light, because of the way they were stimulated, the light waves seemed to move together, synchronized in their ripples. Because of the mechanism, amplifying the initial weak source of microwave photons, it was described as microwave amplification by the stimulated emission of radiation – a maser for short.

By 1960, the American Theodore Harold Maiman had developed an equivalent device for visible light. The concept had been the subject of a patent battle between by American physicist Arthur Leonard Schawlow and another American, Gordon Gould. Gould was eventually recognized as the theoretical originator of the visible maser that Maiman was to build. Gould called his concept a laser, replacing the 'microwave' in maser with 'light'.

Unlike the maser, Maiman's device used a solid substance to produce the stimulated emission – a ruby, giving out a deep red light. The light was stimulated using a flash tube like a huge photographic flash unit. Inside the ruby, the light passed backwards and forwards, hitting mirrors at either end, each time stimulating more photons. One mirror was only partly silvered, allowing part of the beam to escape while some remained in the system.

Because of the way that laser light is produced it is entirely different from the rays of the Sun or an incandescent bulb.

Thought of as a wave, each wave of the light moves in step. Using Feynman's diagrams, the photons' probability arrows are synchronized. The result is a very powerful beam of light of a single colour that is not easily scattered and dispersed as ordinary light is. A laser beam can be bounced off the Moon and still return as a tight ray.

A web of glass

Two further developments that pre-dated the laser would help its rapid transition from novelty to necessity. One, fibre optics, dates back conceptually to 1854. It was an accidental discovery by the same William Tyndall who had tried to explain the sky's blue colour as a scattering effect of dust. Tyndall was watching water stream from a hole in a tank when he noticed that a dancing spot of light accompanied the splattering of the water on the ground. The stream of water was channelling the light, bouncing it back and forth from the edges of the flow like so many tiny mirrors until it arrived at the floor.

The effect is refraction carried to the extreme. Light travelling through the water and hitting the boundary between water and air is bent towards the water's surface. Eventually, the angle of bending is such that the light runs parallel to the water. At a more extreme angle still, the light bounces back from the edge of the water, never escaping. The effect is called total internal reflection.

Light might always travel in straight lines, but by capturing it in a reflecting surface like the stream of water, it could be made to follow a curve. This was the mechanism of the fibre optic, a channel for light. Within 10 years of Tyndall's tank, Charles Vernon Boys had manage to make the first glass fibres, using a romantic technique of attaching a half-set piece of molten quartz to an arrow and shooting it through the air with a bow, stretching out a hair-thin filament of glass. These delicate light pipes remained a novelty until Maiman's remarkable invention turned them into a vehicle for mass communication.

Solid ghosts

The second development that was made possible by the invention of the laser was the work of Hungarian-born British scientist Dennis Gabor. Soon after the Second World War, Gabor was thinking about the way we see objects. Imagine looking through a glass window at a mug on a table. Stand over to the left and you see a certain view of the mug – perhaps the handle and the front side. Move round to the right and the view gradually changes, taking in different angles of the three-dimensional object. All the light required to make up these different views is falling on the window glass. So if there was some way to take a snapshot of all of it, of every ray of light travelling from the mug to the glass, you should be able to recreate the view from the window, with an image that changes as your viewpoint does.

To cope with all the photons coming from different directions you would need to distinguish not just how bright a particular point is, as an ordinary photograph does, but also what stage of its ripple the wave is at – known as the phase (the equivalent of knowing the sum of Feynman's probability arrows for those photons). To do this, Gabor imagined using a second beam of light falling straight on the glass. The two beams, the one bounced off the mug and the other directed onto the glass, would interfere with each other like the beams passing through

Young's slits. The resultant pattern would indicate just what phase of the wave each bit of light was at when it hit the glass.

When he devised his three-dimensional pictures, Gabor's intention was to improve the electron microscope by enabling it to produce an image that could be seen from a range of directions. Gabor's science was always driven by an immediate practical application. In his teens he had built a sophisticated laboratory at home with his brother George. There they went well beyond building a crystal radio set to constructing X-ray devices and experimenting with radioactivity. With this practical bent, Gabor originally studied engineering rather than physics (reasoning that there were many more jobs available for an engineer), but he attended college in Berlin at a time when great names like Einstein and Planck were active, where his practical drive was tempered by an interest in the underlying science. In the case of his 3D microscope, though, the practical result escaped Gabor.

Even when taking the simpler approach of using light, there was a problem – Gabor couldn't make one of these pictures (they were soon called holograms from the Greek *holos* meaning whole and *grapho* to write), because they would only work if the light came from a special kind of source that didn't exist – a source where all the light waves moved in step. Once the laser was produced in 1960 the theory was all ready to be put into practice, and it only took four years before Emmett Leith and Juris Upatnieks at the University of Michigan produced the first true hologram, a bizarre still life of a model train and a pair of stuffed pigeons.

In the years following the 1960s, lasers became commonplace. And the 1960s would also see Einstein's attempt to topple quantum theory, EPR – quantum entanglement – take the first teetering steps towards reality.

Chapter Ten

Tangled light

God runs electromagnetics on Monday,
Wednesday, and Friday by the wave theory,
and the devil runs it by quantum theory
on Tuesday, Thursday and Saturday.
WILLIAM HENRY BRAGG

Back in 1935, Albert Einstein had thrown down a gauntlet to the champions of quantum theory. He had realized that if quantum theory were true it should be possible for two photons separated to opposite sides of the Universe to influence each other. This was such a blatant contradiction of special relativity that he was sure that his EPR paper on this 'quantum entanglement' effect would spell the end of quantum physics and a return to sanity. Einstein was rarely wrong when it came to science, but this time he was wonderfully far off the mark.

Ist mir Wurst

The original paper in which Einstein, Podolsky and Rosen came up with the concept was convoluted. Einstein later commented that it was unnecessarily complicated and poorly written. It involved two entangled particles shooting off in different directions. The imaginary scientists involved (this was only ever intended as a thought experiment) would then make measurements of the position of one particle. Instantly the other's position should be fixed at the same value in the opposite direction, even though up to that point it only existed as a range of probabilities. In the same way, the paper went on, you could measure the momentum (mass times velocity) of one particle, and doing so would fix the momentum of the other, however far away it was. All this measuring of different quantities caused confusion. Einstein later wrote of the need to take both measurements '*ist mir Wurst*', a German phrase which translates literally as 'is sausage to me', but carries the meaning 'I couldn't care less about it'.

The concept of quantum entanglement is much simpler than the paper showed it to be. Once two photons are entangled, it's almost as if these photons become part of the same entity. Separate them as far as you like, and a change in one is instantly reflected in the other. Instantly. The reason this so offended Einstein's sense of propriety is that it seemed to involve 'action at a distance', a concept that horrifies scientists because it seems more akin to magic than science.

If I want to make something happen at a distance, I have to send something from where I am to the place I want to make a change. I might be throwing a ball at a coconut on a fairground stall to knock it over. I might be sending sound waves through the air to activate someone's hearing. Or shining a light to send a signal. In each case, something passes from one place to another. Even gravity, for so long appearing to be action at a distance, is now thought to involve gravitons, gravitational particles, moving from one body to another at the speed of light.

Since Einstein had already established that nothing can travel faster than light, he was confident that entanglement couldn't exist, because that would imply the two photons that were entangled communicating with each other instantly. There's no time for any communication to get from one photon to the other. If nothing crosses the space, then either it's the uncomfortable action at a distance, or it doesn't work, and with it collapses quantum theory.

Making entanglement real

At the time, Einstein's paper caused a brief fuss but was swept under the carpet. Because it was over-complex, most of quantum theory's supporters just pointed out the flaws in the way the paper was written without bothering to worry too much about the fundamental issue of action at a distance. Apart from anything else, the experiment, like many of Einstein's thought experiments, was impractical to carry out. It was almost forgotten for around 30 years, until John Bell, a physicist from Northern Ireland, came up with a different way of examining entanglement that would, in principle, enable a statistical measurement to be taken to demonstrate one way or another whether or not entanglement worked – and hence, whether or not quantum theory was wrong.

Bell's paper passed by pretty well unnoticed. Quantum theory was so well established that there didn't seem much else to be done in the field, and most physicists were concentrating on other more fashionable topics. It took an unconventional Frenchman, doing voluntary work in Africa, to see that it was possible to turn Bell's theory into a practical experiment. Alain Aspect had taken time off in 1971 to live as an aid worker in the central African country of Cameroon. In the long evenings, with little to do, he had time to mull over scientific issues that interested him, rather than concentrating on whatever was fashionable with the physics hierarchy; this included John Bell's paper. When he returned to Paris, Aspect was ready to take on entanglement. He managed to show experimentally that a change made in one photon was reflected in the other half of an entangled pair before there was time for the information to get from one to the other.

To produce entangled photons in the first place, Aspect used calcium, raised to an intense heat. The high-energy calcium atoms were blasted with a pair of lasers, which knocked some of the atoms up into a higher energy state. Soon after, the atoms would drop back down, giving off a photon of light. This is much the same process as happens in ordinary reflection of light. But every once in a while, instead of one photon, two were given off, each of lower energy. These two photons were 'born' entangled, the optical equivalent of twins.

Aspect's technique to generate entangled photons was very inefficient. Since then, two other approaches have taken over. The first, grandly entitled 'parametric down conversion', uses a similar approach to Aspect's but instead of getting 'twinned' photons out of hot calcium, they come from lasers shone through special crystals in a much more controlled fashion. The alternative approach, which can be used with any quantum particle, not just photons, involves the use of beam splitters.

A beam splitter sounds like a prop on an old science fiction TV show, but it's a very common item in the optics lab. Not only there – you have plenty of beam splitters at home, though they were probably designed with another task in mind. We've already seen (see page 236) how a window reflects some of the photons that hit it and lets others pass through. This is a beam splitter in action. The splitter takes a stream of quantum particles, reflecting some and letting others through.

Remember how the thickness of the glass influences the number of photons bouncing off both surfaces of the glass. A photon hitting the inside surface of a window somehow knows how thick the glass is and acts accordingly. There's something similar to entanglement happening here, a kind of action at a distance by which the photon knows how thick the glass is without passing through it, so it doesn't seem entirely surprising that there's a way of shooting two photons through beam splitters and getting them into an entangled state. The exact mechanism is rather messy, but the principle that a pair of beam splitters can get particles into entanglement is one that has been shown to work effectively in entangling everything from photons to clouds of gas.

A message from the future

It isn't only the means of generating entangled photons that has moved on. So have the applications of quantum entanglement. During the 1990s and the early part of the twenty-first century, three major uses of entanglement have emerged. But before looking at these, it is worth considering the most obvious possible use of entanglement. Most people, when they first hear about this strange connection, think that here at last is a way of sending information faster than light – after all, the two entangled particles communicate instantly. As we've seen, faster than light information is a frightening thought, as the message travels backwards in time and can put the whole structure of cause and effect at risk.

This instant communication can travel against the time stream thanks to the way that light's constant speed influences relativity. Two events that would be simultaneous if the locations where the events take place were motionless with respect to each other shift in time as the locations move. If one event comes before another in time, then by the time their relative motion is the speed of light, the order in which the events take place crosses over. Effect comes before cause. Imagine we have a spaceship travelling away from the Earth for 20 years. Because it is moving at a considerable percentage of the speed of light, the travellers only see ten years have elapsed. So an instant message from Earth to the ship will arrive ten years before it was sent. But the effect is symmetrical. From the ship's viewpoint, time is running slowly on the Earth. So in the ten years elapsed on the ship's clock, it sees only five have passed on Earth. The instant message gets back to Earth fifteen years into the past.

If it truly were possible to send time signals, you might wonder why we haven't got messages piling in from the future. But this kind of information time machine has an inescapable limitation. It took our imaginary probe 20 years to reach a distance where it could send a message back fifteen years in time. Even if the probe were to go as close as possible to the speed of light, the message could never get back to Earth at a time before the probe was launched. We can't send messages back before the technology to produce an instant message has been put into operation – and that technology isn't here yet.

Reassuringly, perhaps, quantum entanglement can never help us to build such an information time machine. Although the link between the entangled particles is instant, we can't control what the outcome of that communication is. If we measure the polarization of one of the entangled photons, for instance, it can come out 'up' or 'down'. Instantly, the other entangled particle will click into the opposite state. But we didn't know which way the original photon would go. There was no way to force it to be up or down. So entanglement couldn't be used to send any information.

Another possibility would be to have a long row of entangled photon pairs. The sender looks at some of the photons. When you examine a photon, the entanglement collapses. All you have to do then is check the receiver's photons to see which are still entangled – again you should have instant communication. However, the only way to check for entanglement is to bring the two photons back together – and that can't be done faster than the speed of light. You will have to wait until a boring old light-speed photon has been sent from sender to receiver before you can say whether or not the two particles were still entangled. Whichever way you try to get round it, it is impossible to use entanglement to send a faster than light message.

Entangled secrets

Just because we can't communicate faster than light with entanglement doesn't rule it out from having an important role to play in sending messages. It has proved to be a superb way to keep information secret. A device using quantum entanglement has already been used to send a secure payment from the City Hall in Vienna to a local branch of the Bank of Austria, while Singapore has plans to build a country-wide quantum entanglement network for secure communication. The reason entanglement is so attractive is that it provides a practical way to generate an unbreakable code. (Purists would point out the approach used is really a cipher, which involves replacing individual letters, rather than a code, where special words stand in for whatever you specify they mean, but 'code' seems a more natural word to use.)

Readers of Dan Brown's thriller *Digital Fortress* might be surprised to discover that unbreakable codes have been around since 1918. The plot of Brown's book depends on the shocking discovery that someone has produced unbreakable encryption, which the cipher expert in the book solemnly announces is impossible. No one told this to Gilbert Sanford Vernam, an engineer at AT&T, who suggested keeping information safe by combining each letter of the text to be encrypted with another different value, known as the key. The same key is then subtracted by the recipient to read the message. With the modification dreamed up by Captain Joseph Mauborgne of the US Army Signal Corps, that this key should be a random string of characters, a totally unbreakable mechanism called the one-time pad cipher was born.

This type of code is impossible to break because the enciphered message is itself a set of random characters. There is no pattern, nothing that can be used by a code breaker to untangle it. The only way to get to the information is if you have the key. Yet strangely, even though this technique has been available since 1918, it isn't often used. In the Second World War, for instance, the Germans used Enigma machines – mechanical devices that generated a breakable code, even though it was thought to be impossibly difficult, which is why the code breakers at Bletchley Park were able to crack it. Similarly, the encryption used by computers – for example when you enter your credit card number on a web page – is breakable. Admittedly it is pretty well impossibly difficult (at least for today's technology), but it is breakable. And this despite there being a way to keep secret information totally undecipherable.

The reason the one-time pad isn't used more often is that to make it work you have to get the key to both ends of the link. Both sender and recipient need to have a copy of the list of random characters used to encrypt the information. And getting that key safely from one place to the other (not to mention keeping it safe from prying eyes at both ends of the conversation) means that the onetime pad is often more trouble than it is worth.

Enter quantum entanglement. Imagine a transmitter sending out two streams of entangled photons. One half of each pair goes to the person who wants to send a secret message, the other half goes to the receiver. The key, the random set of information, is generated automatically by the entangled pairs. If, for instance, the encryption device measures polarization, then the sequence of 'up's and 'down's measured for photon after photon is truly random – not fairly random, like the so-called random numbers generated by a computer – but absolutely random. Even better, the key didn't exist before the photons were examined, so there is no way it could have leaked out beforehand. And to crown it all, if anyone eavesdrops on the message, trying to intercept the key, the very action of looking at the photons collapses the entanglement. Provided the two ends of the conversation exchange a steady stream of information that checks on the continued entanglement, they will be instantly alerted if someone intercepts their photons, before any secrets are lost.

Bits and bytes of light

A second, and even more dramatic, use of quantum entanglement is rather further down the technology pipeline, but has already been demonstrated on a very small scale. This is quantum computing. When they are eventually built, quantum

computers will use quantum particles – photons, for example
– instead of bits on a silicon chip to do their computation.
In principle, a quantum bit (called a qubit, pronounced cue-
bit, for short) can process infinitely long numbers. Where an
ordinary bit can only deal with 0 or 1, the settings of the qubit
are its quantum properties, like its spin or polarization. The
polarization of a photon can be in any direction at all around
the axis of its movement. To represent it exactly as a number
would take an infinitely long decimal number. It's pretty obvious such quantum computers should be able to do amazing
things. But there is a huge problem attached to getting quantum computers to work.

Although the information is there in a qubit, it's very difficult to get anything into it or out of it. If you measure the
polarization, you only get one of two values, parallel to the
direction measured or at ninety degrees to it. The *probability*
with which you get parallel or perpendicular is what we think
of as the direction of the polarization – this has an infinitely
long decimal value. But we can't measure that, just parallel or
perpendicular. It's a bit like watching a game of pool on a black
and white TV. All the information is in the pool hall in the real
world, but you can't get to it through the screen; you can only
see the indistinguishable grey balls. Those who are working on
quantum computers – and there are many teams around the
world – all agree that the only way to get information to flow
in a quantum computer is to use entanglement. They simply
couldn't be built without it.

If a full-scale quantum computer can be built, we already
know some of the things it will be able to do, because (perhaps
surprisingly) some of the programs to run on such computers
have effectively already been written. Just a couple of examples

of these will show how much more a quantum computer can do than the everyday variety.

Needles in a quantum haystack

The first case of quantum computing brilliance involves a quantum version of the search for a needle in a haystack (the original paper describing this method was titled 'Quantum mechanics helps in searching for a needle in a haystack'). The algorithm (a set of mathematical rules that in this case can only be used on a quantum computer) was devised by Lov Grover at Bell Labs and provides a way to speed up unstructured searches immensely. Anyone who has used a phone book knows that it's easy to look up a number if you know someone's name. That's a structured search, because the phone book is listed in alphabetical order of names. But try finding who a particular telephone number belongs to. Then the task becomes much more difficult. Imagine you had a phone book with 1 million listings in it. You might have to look at 999,999 entries before getting to the number you wanted. On average you would have to check 500,000 before finding the right number.

With Lov Grover's quantum algorithm, you only need to look at the square root of the number of entries – in this case 1,000 – to be sure of finding what you want. This amazing speed-up of searching will become more and more essential as we deal with the complex messes of information that our increasingly connected world presents us with. There comes a point when any conventional computer will take too long to do a search through unstructured data. But a quantum computer could do it in the square root of the expected time. Variants on the same method allow complex routing problems beyond the capabilities of Google Maps or GPS software to be handled easily. The way route planning software gets to the 'best' route is always an approximation. It's impossible for a conventional computer to come to an absolute solution of a complex problem of this sort in the lifetime of the Universe – but quantum computers would find it trivial.

The quantum code breaker

Another application of quantum computing that already has the algorithm waiting for the hardware to run it on to be built has computer security experts trembling. This is the ability to break down a huge number into the two prime numbers that have been multiplied together to generate it. With big enough numbers, this problem is beyond the capability of any conventional computer we can envisage. But a quantum algorithm already exists to crack the problem, if only there were a quantum computer to run it on. Why should we care? Because the encryption used on every computer – for example when you see a little padlock to tell you that you're safe when you enter your credit card number into a web browser – is based on a technique that relies on the difficulty of breaking down huge numbers into a pair of primes. If you can work out the primes involved, you can crack the code. Most current computer security would fall apart.

This isn't the only application of this ability to deal with primes, but it emphasizes the frightening potential of the quantum computer. At a stroke, if a quantum computer could be built, it would be able to solve a problem that the whole IT industry assumes is insoluble.

Cloning light

Though quantum computing is dramatic, it still isn't the most amazing application of entanglement. This final possibility comes from the special communication between the entangled photons which enables them to get around a fundamental rule of quantum science – the 'no cloning' theorem. This sounds like a mathematical denial of Dolly the sheep and other cloning experiments, but there is a big (or rather very small) difference between a biological clone and a quantum clone. A biological clone is an animal that starts from the same combination of DNA as another animal.

Two clones start off as pretty well the same initial cell, genetically indistinguishable. But in quantum terms a cell is still huge, with plenty of scope for differences, and over time, both tiny random modifications of the genetic makeup and the influence of the environment mean that clones are not exactly the same. You can check this out, because there are quite a number of human clones out there. Not the ones that strange sects and dubious doctors claim to have bred, but perfectly natural clones – identical twins. Anyone who knows identical twins well will tell you that, particularly by the time they are adults, they are clearly distinguishable. There will be visual differences and personality differences. They aren't truly identical.

When scientists say you can't clone a quantum particle like a photon or an atom, they aren't thinking of this sort of approximate copy. They mean you can't produce an *exact* copy with all the properties of the particles – their spins, for example – set to exactly the same value. The sheer act of looking at a particle to check its spin changes it; the no cloning theorem goes further and proves that it is impossible to take an exact copy of a particle. With quantum entanglement on hand, though, physicists can do the next best thing.

It might not be feasible to make an exact copy of a particle, so that after applying some magic cloning process you have two particles that are exactly the same, but by interacting a particle with one half of an entangled pair, then using some information that comes out of the interaction to make a change to the second half of the entangled pair, it *is* possible for that second entangled particle to be made exactly as the original particle was. In the process the original particle is mangled. It no longer has the properties it started with. And we never find out what those properties were. But the entanglement provides a route for the properties to be transferred to the second particle. This process, known as quantum teleportation, is already being used in prototype quantum computers, and could have even more remarkable uses, which will be explored in the next chapter.

With entanglement the truly remarkable possibilities of using light in new ways become apparent. Although largely confined to the lab, entanglement is already being used for practical purposes. And this is only the beginning of the light revolution.

Chapter Eleven

Tyger! Tyger!

No other science has so much sweetness
and beauty of utility.
ROGER BACON

One after another, the twentieth century iconoclasts ripped apart the carefully constructed classical picture of light. First to go was the ether. Then Young's thoroughbred light wave was replaced by a mongrel, neither wave nor particle, but somehow both simultaneously. And QED made it unnecessary for light to be a wave at all. As Feynman's theories changed our understanding of light's life cycle, light technology was beginning to change our lives.

If the only impact of quantum theory had been to explain the nature of light it still would have been revolutionary. Light had proved to be more than the source of energy for life, filling all matter with a dancing web of photons. But now it has become the driving force behind remarkable products. Quantum light technology lies behind new forms of light, and is enabling scientists to slow light to a crawl and even to capture it. It is now even possible to drive light beyond Einstein's apparently unassailable speed limit. Quantum light devices make the impossible an everyday occurrence.

The remarkable laser

The starting point is the laser. Lasers already crop up in practically every home, in CD and DVD players and recorders, and in printers. They are in action unseen under our streets and oceans, pumping information through the flimsy threads of fibre optics. The power of lasers is only just becoming obvious. Soon they could transform the heart of computers in a more practical way for everyday products than the quantum computer, which is always likely to be a sensitive laboratory beast.

Computing speed is limited by the time taken for the relatively sluggish electrons to make their way around the network of wires and circuit board paths inside the computer box. With the help of lasers, this wiring could be replaced by free space optics. Instead of sending a piece of information down a wire it is beamed across the box on a laser. Not only does the information travel at light speed, there's a double bonus. It's often the sheer space taken up by wires and components that forces a computer to be a particular size. Free space optics can cross each other in space. Photons of light don't interfere with each other. (It's just as well or the mesh of visible light, radio, TV, mobile phone and other signals that are crossing the room in front of your nose as you read this would collide and cause chaos.) By weaving a basket of light inside a computer, the space taken up by the electronics can be significantly reduced.

But there's an even more important change coming to information technology, one with the potential to transform publishing, the Internet and any information-based business. We're used to holograms as clever trick pictures, or in the shiny, multicoloured security stickers on easily copied items, but this is only scratching the surface of what holographic technology can do. Let's travel forward just a few years.

On the leather top of an antique desk are half a dozen crystals, rectangular in shape. They're small enough to fit into the palm of your hand. The woman sitting at the desk picks one up. The crystal's top surface has an embedded gold computer chip that glints for a moment as it catches the light. Beneath, the material is like a smoky glass with swirls and ripples embedded inside the smooth outer skin. She holds it up and peers through. It has the look of a miniature Milky Way, glistening with points of light as if a million tiny stars have been captured inside. It could be a child's toy, but combined with a beam of laser light, this crystal and its fellows have devastated the information industries.

This is more powerful than any secret weapon or industrial espionage. The crystal is frozen light. A laser has modified the lattice of molecules inside the crystal, changing it from a simple repeating pattern to the complex swirls of a three-dimensional hologram. Packed in the tiny space is a vast amount of information. That handful of crystals on the desk holds every single book in print in the English language. These crystals, which will turn publishing, computing, the Internet on their heads are being fine-tuned in the laboratory today. Their power lies in information on an unparalleled scale. It's like the move from a few dozen laboriously hand-copied manuscripts to a million books in print. Or from a single encyclopedia to the whole of the Internet. The frozen light of the holographic crystal will transform our information hungry world.

Whirlpools of light

Although the freezing of light in holographic crystals is only metaphorical, another technology currently under development involves bringing light to a complete standstill. It is the next step on from the slow glass experiments of Lene Vestergaard Hau described in the first chapter. Once again it's Einstein's fifth state of matter, the unique Bose–Einstein condensate, that is involved.

The intention is to pull photons into spinning vortexes in a Bose–Einstein condensate, hoping that the light will be dragged into the churning matter like a car sucked into a tornado. If these frigid whirlpools can be spun fast enough, they will become microscopic optical black holes, clawing in light and never letting go until the vortex loses its momentum. Ulf Leonhardt and Paul Piwnicki of the Stockholm Royal Institute of Technology and the University of St Andrews in Scotland believe that such an effect can be produced.

These optical black holes are made possible by the extreme slowness of the light passing through the condensate. Unlikely though it sounds, the light moves so slowly that it is entirely feasible to spin the vortex of condensate faster than light moves within it. This won't result in any distortions of time – the effects of Einstein's special relativity depend on moving at close to light's ultimate speed in a vacuum, not its slower speed in a material. But the extremely slow velocity of light in the condensate does mean that the medium will be moving faster than the light it carries, enabling the rotating atoms to drag light photons with them into the vortex, never to escape until the spin slows down.

It might seem for a moment as if special relativity has gone wrong. After all, the theory says that light's speed should be the same whatever the relative movements around it. But there are two effects occurring here – the simple motion of the condensate and the way that the photon's electromagnetic makeup interacts with the electric field of this special moving matter. It's the second of these that is the key. The vortex's ability to stop the light is not a simple matter of relative motion, but the effect of creating a spinning electrical field at faster than local light speed.

This approach has had some initial success in experiments, though it has yet to be fully explored. At the same time, Lene Hau's team have not stood still since they originally slowed light to a crawl, despite accidental sabotage by a German TV team. The strange possibilities of quantum light experiments quite often attract media attention, but a modern lab is visually boring. One set of black boxes looks much like any other. The TV team decided that they could make Hau's experiments look more impressive by bringing in a smoke machine to make the interlacing patterns of lasers visible. Unfortunately they didn't ask permission to do this. The result was a total collapse of the experiment, which had to be shut down for days until the air could be cleared. Now a plastic curtain surrounds the table that houses the experiment to keep out interfering onlookers.

As we saw in the first chapter, Hau's first experiments used one laser to form a sort of ladder through the otherwise opaque Bose–Einstein condensate that allowed a second laser to claw its way through. But if that first laser, called the coupling laser, is gradually decreased in power, the team found that the second beam was swallowed up in the material. The result is a strange mix of matter and light, called a dark state. The trapped light only comes out again when the coupling laser is restarted.

Developing optical black holes or dark states into a practical product will take more than clever manufacturing. You've got to understand light at a quantum level, working with the individual photons, the tiny packets of energy that make up the light beam. And quantum physics will always be mind-bending because there are no absolute certainties. Everything depends on probability. Remember that entanglement came out of Einstein's failed attempt with Podolsky and Rosen to show how ridiculous quantum theory was. We have already seen some potential applications of this remarkable phenomenon, but quantum teleportation may go even further in the future.

Beam it up

In that laboratory of the future, the scene is likely to be as much a media event as a scientific breakthrough. Imagine a workbench, crowded round by a bizarre mix of scientists and celebrities and media presenters. In front of them, at the centre of a tangle of wires and pipes, is a small transparent chamber, the size of a fist. Those lucky enough to be at the front feel the edge of the bench cutting into their stomachs as everyone else struggles to get a better view. There's a faint smell of oil and electricity in the air.

As the antique analogue clock on the wall ticks towards the hour, the jostling stops. It's as if everyone has forgotten to breathe. It's so quiet now that they can hear the clock's even beat. Within the chamber the view becomes distorted, like the churning air over a searing hot desert road. And now, inside the sealed-off space, a small child's toy appears. A building brick, painted bright red with a worn picture of a panda on the side. There is a last moment of silence before the cheering begins. A few seconds later and the yells spread to the screens linking them with Beijing. In an instant the child's brick has been moved from one side of the world to the other. Solid matter has taken a leap through space, as if propelled by the transporters of *Star Trek*'s USS *Enterprise*.

Total fiction? Perhaps not. Since 1997, when quantum teleportation was first achieved in a fragile way, transporting a single particle across a laboratory, big steps have been taken in making the process more robust. In 2004, Anton Zeilinger, one of the first to use teleportation in 1997 and the leading light in the entanglement field, succeeded in teleporting a particle across the river Danube. Zeilinger has also managed to get much larger quantum particles than photons into the superposed state required to use a beam splitter to entangle them – all the way up to molecules similar in size to a virus – though a physical object of the size of the building brick in my fictional example would be impossible to get into a superposed state. An object this size would need to be stripped down, scanned and rebuilt at the other end.

In principle this process could be used to transmit each particle making up a solid object, though the technology to rebuild the final object does not yet exist – but it is a technological problem, not a theoretical one. Anton Zeilinger is doubtful about going much further than a molecule, though. 'Nothing in principle limits [the size of the object that can be teleported]. For sufficiently large objects – probably anything living – teleportation is still a fantasy, but you never know!' As Zeilinger himself has commented, 'an experimentalist should never use the word "never". Some of the experiments we are doing today I would never have believed possible ten years ago'.

Interestingly, the need to send the tangled photons and the information by conventional means makes this approach unsuitable for travelling stellar distances (something that, by accident, *Star Trek* got right). It also makes quantum teleportation an impossible approach for time travel – although the spooky interaction does travel faster than light, and hence backwards in time, by the time the whole process is undertaken it has been significantly slower than light speed.

Teleportation may be possible in theory, but should a matter transmitter be constructed it will present a frightening ethical issue. Although described as a transmitter, the quantum teleportation effect actually functions as a matter duplicator, producing an indistinguishable remote copy. In the process, the original would have to be destroyed as the particle-by-particle spooky linkages are built. It would take a brave traveller to knowingly undertake a process that would totally destroy his or her body, even if an exact duplicate was about to be constructed elsewhere. It's likely that human teleportation, at least, will remain the property of science fiction.

The final barrier

Despite the dramatic developments in quantum-driven light technology, one assumption was reinforced by the new science. The speed of light had always been considered the fastest thing around. Einstein not only confirmed this, but established it as an apparently unbreakable barrier. Exceed the speed of light and time would actually move backwards.

Thanks to Einstein's theory, it's safe to say that nothing solid can ever reach light speed. However tiny the object, it will become heavier and heavier as it nears that velocity. The amount of effort required to move it will grow. All the energy in the Universe wouldn't be enough to get it moving fast enough. But light itself has no choice but to travel this quickly, and with special techniques it can even be pushed past the limit. In recent experiments, light was routinely travelling at over four times its normal speed. Moving at this remarkable rate, the experimenters' pulses of light would not be flowing along normally in time. Instead they would be slipping back against the time stream like salmon fighting their way up a river.

It's one thing to send a simple flash of light so quickly, but if a signal, anything that could carry a message, travelled along such a faster-than-light beam it would have frightening consequences. Pushed far enough back in time, it would arrive before it was sent. As we have seen, it could predict the lottery results, prevent a disaster occurring, make new billionaires and destroy existing ones. The idea of changing the past, of disrupting the cast-iron link of cause and effect, is so mind-bending that it has always been assumed to be impossible. And yet, Feynman showed that light's relationship with time was anything but common sense. Quantum electrodynamics treats time as just another dimension, and has no trouble handling photons that travel backward in time. The stage was set for the shattering of the last and greatest barrier.

The undersized waveguide

Great beginnings often come in strange places. Ideas, reluctant to appear when the thinker is sitting at a desk, have a habit of popping up on a walk, in the car or at the gym, unannounced and surprising. It's a side-effect of the way the brain operates. When directed and focused it follows well-trodden paths. When meandering and daydreaming it is much more likely to make the new connections and linkages that are necessary to spark an idea. It's no surprise that Einstein's own decisive thought experiment at the heart of relativity took place while lying on a grassy bank.

Professor Günter Nimtz of the University of Cologne in Germany was returning on the train from a meeting in Stuttgart. The scenery was uninspiring; the meeting left Nimtz with little to challenge his imagination. He began to read through a paper published by Doctor Anedio Ranfagni and his colleagues at the National Institute for Research into Electromagnetic Waves in Florence. The Italians' experiment involved pushing light through an undersized waveguide. Like John Tyndall's water pouring from the tank, a waveguide is a conduit that carries light by total internal reflection. Outside the visible part of the light spectrum, as in this experiment, waveguides tend to be rectangular metallic tubes. Of itself there was nothing unusual in the waveguide experiment, but there was something odd about the results. Nimtz frowned.

He read it over again, then showed it to his postdoctoral student, Achim Enders (now a professor at the University of Braunschweig), who was travelling with him. By the time Nimtz and Enders had reached Cologne they were determined to reproduce the experiment. The Italians claimed that the effect of pushing light down this undersized pipe was, from a mathematical viewpoint, the same as getting it to tunnel through a barrier, taking the special quantum-level jump through a solid object described in Chapter 1. This, thought Nimtz, was fair enough. But the light appeared to travel at much less than its usual speed. It was this slowness that made Nimtz suspicious.

When Professor Nimtz saw that tunnelling appeared to slow down the light beam, it seemed intuitively wrong. The Italian paper acted as an irritant to his curiosity, like a grain of sand stimulating an oyster into producing a pearl. Until then, his main concern had been electromagnetic shielding, investigating materials that act as barriers to electromagnetic radiation above the visible spectrum. This was more than a matter of curiosity. The Tornado fighter/bomber's electronic control systems were proving dangerously susceptible to disruption by wild light, the wide-ranging electromagnetic discharges generated by the plane's engines and other sources. Nimtz continued to work on shielding, but couldn't resist the intellectual challenge of the tunnelling problem that was to trigger some remarkable developments.

When Nimtz's team in Cologne reproduced the Italian experiment they found very different results. Their measuring equipment was better suited to the problem, and with some fundamental errors overcome it became obvious that the tunnelling did not slow down the light, but made it faster. The undersized waveguide pushed light beyond the normal 300,000 kilometres a second and hence raised the possibility of challenging Einstein. Sometimes, such alternative findings might result in a battle between the two groups, but here there was little doubt, and before long the Italian team accepted their error.

At around the same time, a group at the University of California at Berkeley headed up by Professor Raymond Chiao was achieving similar results from a more conventional barrier. Here visible light was pushed through a photonic lattice – a sandwich of layers of material of a very different refractive index that stops photons of a particular frequency from passing through. Some photons manage to penetrate the obstacle by tunnelling, and these appeared to arrive at nearly twice the speed of light. Chiao didn't find this disturbing. Because the technique he used to generate appropriate photons for his experiment meant that they were produced at random, there was no way to send a signal, which meant no possibility of getting information to travel back in time and disrupt causality. The next year, though, Günter Nimtz, came up with another experiment that seemed to show something very different, and being Nimtz he publicized it in a very dramatic way.

Mozart 40

It was January 1995, in the ski resort of Snowbird, Utah, perched at the top of Little Cottonwood Canyon, 8,000 feet above sea level. Günter Nimtz was attending a session on superluminal velocities – faster than light speeds – at the Optical Society of America's annual meeting.

When it was time for Nimtz to speak, he took a Walkman from his pocket and carried it to the front of the room. He put on half-moon glasses, peered down at the notes, and then began to talk, walking back and forward in front of the delegates. His English was good, only betrayed as a foreign language by the occasional hesitation as he searched for the right word. Unlike his colleagues he used very few equations, relying instead on graphs that he slipped onto the overhead projector like a magician pulling rabbits from a hat. Some of the delegates were struggling to stay awake after a long evening, but suddenly Nimtz captured their attention.

'Our colleagues assure us that their experiments do not endanger causality, that there is no possibility of sending a message into the past.' He paused for a wry smile, directed straight at Raymond Chiao. Chiao's face was impassive.

Nimtz continued. 'They are happy that no signal can be

transmitted faster than light. But I would like you to listen to something.' He straightened his son's much-handled Walkman, aligning it with the edge of the table in front of him, before pressing the play button.

From the speaker came a hissing, and then, faintly but clearly, a dancing sequence of musical notes, the opening of Mozart's 40th symphony. Nimtz allowed the music to echo tinnily around the room for a few moments until the woodwind and horns emphatically reinforced the strings. 'This Mozart', said Nimtz, 'has travelled at over four times the speed of light. I think that you would accept that it forms a signal. A signal that moves back in time'.

Nimtz's enthusiasm for showmanship inevitably irritated some of his colleagues, making them less inclined to treat his experiments seriously. And there is also a possibility that his background in engineering before moving into physics was responsible for some of this suspicion. Scientists have always tended to hold the more practically minded engineers in disdain. The rivalry between the different disciplines is illustrated by a joke popular with mathematicians and physicists.

A mathematician, a physicist and an engineer are trying to work out of if all the odd numbers are prime (can't be divided by another number to produce a whole result). The mathematician quickly counts off: 1, 3, 5, 7, 9 – no, nine isn't prime, it can be divided by three, so it's not true. The physicist does much the same: 1, 3, 5, 7, 9 – hmm, 11, 13 – yes, he says, odd numbers are prime, 9 was just an experimental error (bad scientists have a habit of ignoring data that doesn't fit their needs). Then it's the engineer's turn: 1, er, er, 3, erm…

It's also true that Nimtz takes a positive delight in teasing some of his more straight-laced colleagues. Shortly before his

Mozart demonstration, a pair of respected US researchers had concluded that it wasn't possible to send information faster than light. Nimtz countered by commenting that 'maybe to an American, Mozart's 40th isn't information', a jibe he would later turn against Raymond Chiao in a BBC TV show on time travel. Chiao's response was enigmatic; the other American researchers were less gracious, writing Nimtz a letter accusing him of arrogance. When the eminent physicist Francis Low saw a demonstration of Nimtz's party piece his immediate response was to say nothing. He walked up and down for at least a minute before commenting 'that's not G minor'. Low, known to have perfect pitch, was happy to comment on the quality of the recording if not the experiment.

Nimtz's experiment involved transmitting the music across space, just as if it were a radio broadcast, but using light in the microwave band. When a barrier was put in place between transmitter and receiver – first an undersized waveguide as used in the Italian experiment, then later a photonic lattice like Chiao's – in each case, the signal had continued, tunnelling through the barrier. The result was weak and distorted, but clearly still Mozart's 40th. And the Cologne group's delicate measuring instruments showed that the signal was arriving earlier with the barrier in place – it was crossing the gap between transmitter and receiver faster than light speed.

Crossing the finish line

Even though the Mozart demonstration was only ever intended as a provocation, it shows how fine distinctions of language are necessary when as sensitive a possibility as time travel comes up. In fact, Nimtz himself is happy to admit that, while information was transmitted at four times the speed of light, it would not have been possible to gain any time advantage by using this technique. To understand this apparent paradox, you have to look closely at the nature of both light and a signal.

One of the problems with light that makes many physicists doubt that information is really managing superluminal speeds is that light doesn't have a single, simple velocity. Imagine a pulse of light, a very short blip. You could say that the speed of the light was the speed with which the most intense point of the blip moved forward. Or you could say it was the speed of one of the actual light waves within the blip. In many circumstances, these are the same, but not always. The differences between the two can distort the shape of the blip during its travel, making it appear to have travelled faster than it really did.

Imagine two runners in a race. Say that the winner is the one whose hands cross the finishing line first. Both start at exactly the same time and run at the same speed, but one sticks his arms out, the other keeps them by his side. Despite running at the same speed, the runner with his arms out will arrive a little earlier, and hence seems to have run faster. What has actually happened is his shape has changed. Similarly a light pulse can be reshaped when it passes through a barrier, producing a misleading result.

Nimtz got round this problem by using light that was limited in its bandwidth – the range of frequencies present. This prevents reshaping and the possible ensuing confusion. However, he also points out that we need to remember just what a signal *is* to understand why the world is not about to come unravelled. At its heart, a signal is a series of 0s and 1s, like the bits in a computer. This is sent along a light beam (whether to your car radio or TV or to the receiver in Nimtz's experiments) by a process known as frequency modulation. The signal starts as a 'carrier', a smooth, steady wave. The information is then added to the wave, so that by making the next up and down motion come a little sooner (say) a 1 is indicated.

However, we can't tell whether a 0 or a 1 is being sent until the wave has completed its up and down motion once. To actually gain a march on time, the wave needs to get ahead of time by one whole up and down motion – and that has not been achieved. All the experiments have managed is a small percentage shift against the wave itself. Mozart's 40th was shifted in time – but only by a fraction of a wavelength.

Nimtz's natural showmanship and practical bent turned up again in a later experiment, in 2000. This uses a strange phenomenon that was hinted at by Newton – frustrated total internal reflection. If two prisms are placed face to face with a gap between them that is filled with a less refracting substance, there are angles where a small part of the light that should be reflected in fact comes out through the second prism. In doing so it tunnels across the gap, and can be easily demonstrated to cross in no time. Nimtz uses huge plexiglass prisms, 40 centimetres to a side, and microwaves to demonstrate this, obviously as delighted by the beautiful simplicity of the apparatus as the results.

Most dramatically of all, 2000 also saw a group headed up by Doctor Lijun Wang at the NEC Research Institute in Princeton push a pulse of light at what has been described as 310 times the normal speed. In fact, Wang's experiment had an even more remarkable result – the velocity of light was measured at $-1/310$th of its usual speed, effectively arriving before it started with a more than relativistic backward motion in time.

Wang's experiment used a very different approach,

depending on the way that a light beam is changed by passing through a specially prepared tube of caesium gas. By using a similar concept to the way a laser builds up a powerful beam by stimulating new emissions of light, Wang was able to produce a pulse that was already leaving his tube before it had properly entered. The numbers are impressive, but misleading from the point of view of a faster than light signal. Because the pulse has to be relatively wide in time for this experiment, the shift in the signal was a fraction of that of earlier experiments. According to Nimtz it would take an effect more than 100 times more powerful to get a usable backward shift in time by using this technique.

For the moment, such experiments remain solely in the domain of the laboratory. The effect is to tweak at time's skirts without doing anything to upset the fundamental workings of the Universe. But will it always be so? Professor Nimtz is not sure. Like Anton Zeilinger, Nimtz comments 'I never say never'.

Gunter Nimtz sent a signal, Mozart's 40th symphony, as much a set of information as Kennedy's Berlin speech or the blueprints of the Space Shuttle, faster than light. The future of faster than light communication is uncertain, but the attempt is bound to be fascinating.

The tyger unchained

The study of light is the study of a fundamental component of creation. Light is at the heart of matter. It gives us sight, warmth, food and energy on the Earth. It crosses the Universe unchanged, yet is constantly being created and destroyed. This ephemeral yet powerful phenomenon has always been fascinating, but as we move into the twenty-first century, light has become something more than an object of fascination and wonder – it is driving technology that will change our lives, and is even leading us to question the nature of reality.

There is no doubt that light is going to have a fundamental impact on the future. The light revolution is just beginning. Enjoy the ride.

Browsing

In the next few pages there's a chance to explore some of the key documents that have changed our understanding of light. They aren't necessary to understand *Light Years*, but they help give a feel for the approach of past times, when science was very different from today. This collection is in no sense complete – it is a few fragments, a taster to gain a flavour of the roots of light science.

Isaac Newton
On Colour and Light

*In a letter written to Henry Oldenburg, the Secretary
of the Royal Society, Newton gives a description of
his findings when experimenting with light.*

Trin: College Cambridge

Sir

To perform my late promise to you, I shall without further ceremony acquaint you, that in the beginning of the Year 1666 (at
which time I applyed my self to the grinding of Optick glasses
of other figures than Spherical,) I procured me a Triangular
glass-Prisme, to try therewith the celebrated Phaenomena of
Colours. And in order thereto having darkened my chamber,
and made a small hole in my window-shuts, to let in a convenient quantity of the Suns light, I placed my Prisme at its
entrance, that it might be thereby refracted to the opposite wall.
It was at first a very pleasing divertisement, to view the vivid
and intense colours produced thereby; but after a while applying my self to consider them more circumspectly, I became
surprised to see them in an oblong form; which, according to
the received laws of Refraction, I expected should have been
circular.

They were terminated at the sides with streight lines, but
at the ends, the decay of light was so gradual, that it was difficult to determine justly, what was their figure; yet they seemed
semicircular.

Comparing the length of this coloured Spectrum with its

breadth, I found it about five times greater; a disproportion so extravagant, that it excited me to a more then ordinary curiosity of examining, from whence it might proceed. I could scarce think, that the various Thickness of the glass, or the termination with shadow or darkness, could have any Influence on light to Produce such an effect; yet I thought it not amiss to examine first these circumstances, and so tryed, what would happen by transmitting light through parts of the glass of divers thicknesses, or through holes in the window of divers bignesses, or by setting the Prisme without, so that the light might pass through it, and be refracted before it was terminated by the hole: But I found none of those circumstances material. The fashion of the colours was in all these cases the same.

Then I suspected, whether by any unevenness in the glass, or other contingent irregularity, these colours might be thus dilated. And to try this, I took another Prisme like the former, and so placed it, that the light, passing through them both, might be refracted contrary ways, and so by the latter returned into that course, from which the former had diverted it. For, by this means I thought, the regular effects of the first Prisme would be destroyed by the second Prisme, but the irregular ones more augmented, by the multiplicity of refractions. The event was, that the light, which by the first Prisme was diffused into an oblong form, was by the second reduced into an orbicular one with as much regularity, as when it did not at all pass through them. So that, what ever was the cause of that length, 'twas not any contingent irregularity.

I then proceeded to examine more critically, what might be effected by the difference of the incidence of Rays coming from divers parts of the Sun; and to that end, measured the several lines and angles, belonging to the Image. Its distance from the hole or Prisme was 22 foot; its utmost length 13¼ inches; its breadth 2⅝ inches; the diameter of the hole ¼ of an inch; the angle, which the Rays, tending towards the middle of the image, made with those lines, in which they would have proceeded without refraction, 44 deg. 56′. And the vertical Angle of the Prisme, 63 deg. 12′. Also the Refractions on both sides the Prisme, that is, of the Incident, and Emergent Rays, were as near, as I could make them, equal, and consequently about 54 deg. 4′. And the Rays fell perpendicularly upon the wall. Now subducting the diameter of the hole from the length and breadth of the Image, there remains 13 Inches the length, and 2⅜ the breadth, comprehended by those Rays, which passed through the center of the said hole, and consequently the angle at the hole, which that breadth subtended, was about 31′, answerable to the Suns Diameter; but the angle, which its length subtended, was more then five such diameters, namely 2 deg. 49′.

Having made these observations, I first computed from them the refractive power of that glass, and found it measured by the ratio of the sines, 20 to 31. And then, by that ratio, I computed the Refractions of two Rays flowing from opposite parts of the Sun's discus, so as to differ 31′ in their obliquity of Incidence, and found, that the emergent Rays should have comprehended an angle of about 31′, as they did, before they were incident.

But because this computation was founded on the Hypothesis of the proportionality of the sines of Incidence, and Refraction, which though by my own & others Experience I could not imagine to be so erroneous, as to make that Angle but 31′, which in reality was 2 deg. 49′; yet my curiosity caused me again to take my Prisme. And having placed it at my window, as before, I observed, that by turning it a little about its axis to and fro, so as to vary its obliquity to the light, more then by an angle of 4 or 5 degrees, the Colours were not thereby sensibly translated from their place on the wall, and consequently by that variation of Incidence, the quantity of Refraction was not sensibly varied. By this Experiment therefore, as well as by the former computation, it was evident, that the difference of the Incidence of Rays, flowing from divers parts of the Sun, could not make them after decussation diverge at a sensibly greater angle, than that at which they before converged; which being, at most, but about 31 or 32 minutes, there still remained some other cause to be found out, from whence it could be 2 deg. 49'.

Then I began to suspect, whether the Rays, after their trajection through the Prisme, did not move in curve lines, and according to their more or less curvity tend to divers parts of the wall. And it increased my suspition, when I remembred that I had often seen a Tennis-ball, struck with an oblique Racket, describe such a curve line. For, a circular as well as a progressive motion being communicated to it by that streak, its parts on that side, where the motions conspire, must press and beat the contiguous Air more violently than on the other, and there excite a reluctancy and reaction of the Air proportionably greater. And for the same reason, if the Rays of light should possibly be globular bodies, and by their oblique passage out of one medium into another acquire a circulating motion, they ought to feel the greater resistance from the ambient Aether, on that side, where the motions conspire, and thence be continually bowed to the other. But notwithstanding this plausible ground of suspition, when I came to examine it, I could observe no such curvity in them. And besides (which was enough for my purpose) I observed, that the difference betwixt the length of the Image, and diameter of the hole, through which the light was transmitted, was proportionable to their distance.

The gradual removal of these suspitions at length led me to the Experimentum Crucis, which was this : I took two boards, and placed one of them close behind the Prisme at the window, so that the light might pass through a small hole, made in it for that purpose, and fall on the other board, which I placed at about 12 foot distance, having first made a small hole in it also, for some of that Incident light to pass through. Then I placed another Prisme behind this second board, so that the light, trajected through both the boards, might pass through that also, and be again refracted before it arrived at the wall. This done, I took the first Prisme in my hand, and turned it to and fro slowly about its Axis, so much as to make the several parts of the Image, cast on the second board, successively pass through the hole in it, that I might observe to what places on the wall the second Prisme would refract them. And I saw by the variation of those places, that the light, tending to that end of the Image, towards which the refraction of the first Prisme was made, did in the second Prisme suffer a Refraction considerably greater then the light tending to the other end. And so the true caused of the length of that Image was detected to be no other, then that Light consists of Rays differently refrangible, which, without any respect to a difference in their incidence, were, according to their degrees of refrangibility, transmitted towards divers parts of the wall.

When I understood this, I left off my aforesaid Glass-works; for I saw, that the perfection of Telescopes was hitherto limited, not so much for want of glasses truly figured according to the prescriptions of Optick Authors, (which all men have hitherto imagined,) as because that Light it self is a Heterogeneous mixture of differently refrangible Rays. So that, were a glass so exactly figured, as to collect any one sort of rays into one point, it could not collect those also into the same point, which having the same Incidence upon the same Medium are apt to suffer a different refraction. Nay, I wondered, that seeing the difference of refrangibility was so great, as I found it, Telescopes should arrive to that perfection they are now at. For, measuring the refractions in one of my Prismes, I found, that supposing the common sine of Incidence upon one of its planes was 44 parts, the sine of refraction of the utmost Rays on the red end of the Colours, made out of the glass into the Air, would be 68 parts, and the sine of refraction of the utmost rays on the other end, 89 parts: So that the difference is about a 24th or 25th part of the whole refraction. And consequently, the object-glass of any Telescope cannot collect all the rays, which come from one point of an object so as to make them convene at its focus in less room then in a circular space, whose diameter is the 50th part of the Diameter of its Aperture; which is an irregularity, some hundreds of times greater, then a circularly figured Lens, of so small a section as the Object glasses of long Telescopes are, would cause by the unfitness of its figure, were Light uniform.

This made me take Reflections into consideration, and finding them regular, so that the Angle of Reflection of all sorts of Rays was equal to their Angle of Incidence; I understood, that by their mediation Optick instruments might be brought to any degree of perfection imaginable, provided a Reflecting substance could be found, which would polish as finely as Glass, and reflect as much light, as glass transmits, and the art of communicating to it a Parabolick figure be also attained. But these seemed very great difficulties, and I almost thought them insuperable, when I further considered, that every irregularity in a reflecting superficies makes the rays stray 5 or 6 times more out of their due course, than the like irregularities in a refracting one: So that a much greater curiosity would be here requisite, than in figuring glasses for Refraction.

Amidst these thoughts I was forced from Cambridge by the Intervening Plague, and it was more then two years, before I proceeded further. But then having thought on a tender way of polishing, proper for metall, whereby, as I imagined, the figure also would be corrected to the last; I began to try, what might be effected in this kind, and by degrees so far perfected an Instrument (in the essential parts of it like that I sent to London,) by which I could discern Jupiters 4 Concomitants, and shewed them divers times to two others of my acquaintance. I could also discern the Moon-like phase of Venus, but not very distinctly, nor without some niceness in disposing the Instrument.

From that time I was interrupted till this last Autumn, when I made the other. And as that was sensibly better then the first (especially for Day-Objects,) so I doubt not, but they will be still brought to a much greater perfection by their endeavours, who, as you inform me, are taking care about it at London.

I have sometimes thought to make a Microscope, which in like manner should have, instead of an Object-glass, a Reflecting piece of metall. And this I hope they will also take into consideration, For those Instruments seem as capable of improvement as Telescopes, and perhaps more, because but one reflective piece of metall is requisite in them [...]

But to return from this digression, I told you, that Light is not similar, or homogeneal, but consists of difform Rays, some of which are more refrangible than others: So that of those, which are alike incident on the same medium, some shall be more refracted than others, and that not by any virtue of the glass, or other external cause, but from a predisposition, which every particular Ray hath to suffer a Particular degree of Refraction.

I shall now proceed to acquaint you with another more notable difformity in its Rays, wherein the Origin of Colours is infolded. A naturalist would scearce expect to see ye science of those become mathematicall, & yet I dare affirm that there is as much certainty in it as in any other part of Opticks. For what I shall tell concerning them is not an Hypothesis but most rigid consequence, not conjectured by barely inferring 'tis thus because not otherwise or because it satisfies all phenomena (the Philosophers universall Topick,) but evinced by ye mediation of experiments concluding directly & wthout any suspicion of doubt. To continue the historicall narration of these experiments would make a discourse too tedious & confused, & therefore I shall rather lay down the Doctrine first, and then, for its examination, give you an instance or two of the Experiments, as a specimen of the rest.

The Doctrine you will find comprehended and illustrated in the following propositions.

1. As the Rays of light differ in degrees of Refrangibility, so they also differ in their disposition to exhibit this or that particular colour. Colours are not Qualifications of light, derived from Refractions, or Reflections of natural Bodies (as 'tis generally believed,) but Original and connate properties, which in divers Rays are divers. Some Rays are disposed to exhibit a red colour and no other; some a yellow and no other, some' a green and no other, and so of the rest. Nor are there only Rays proper and particular to the more eminent colours, but even to all their intermediate gradations.

2. To the same degree of Refrangibility ever belongs the same colour, and to the same colour ever belongs the same degree of refrangibility. The least Refrangible Rays are all disposed to exhibit a Red colour, and contrarily those Rays, which are disposed to exhibit a Red colour, are all the least refrangible: So the most refrangible Rays are all disposed to exhibit a deep Violet colour, and contrarily those which are apt to exhibit such a violet colour, are all the most Refrangible. And so to all the intermediate colours in a continued series belong intermediate degrees of refrangibility. And this Analogy 'twixt colours, and refrangibility, is very precise and strict; The Rays always either exactly agreeing in both, or proportionally disagreeing in both.

3. The species of colour, and degree of Refrangibility proper to any particular sort of Rays, is not mutable by Refraction, nor by Reflection from natural bodies, nor by any other cause, that I could yet observe. When any one sort of Rays hath been well parted from those of other kinds, it hath afterwards obstinately retained its colour, notwithstanding my utmost endeavours to change it. I have refracted it with Prismes and reflected it with Bodies, which in Day-light were of other colours; I have intercepted it with the coloured film of Air interceding two compressed plates of glass; transmitted it through coloured Mediums, and through Mediums irradiated with other sort of Rays, and diversly terminated it, and yet could never produce any new colour out of it. It would by contracting or dilating become more brisk, or faint, and by the loss of many Rays, in some cases very obscure and dark; but I could never see it changed in specie.

4 Yet seeming transmutations of Colours may be made where there is any mixture of divers sorts of Rays. For in such

mixtures, the component colours appear not, but, by their mutual allaying each other, constitute a midling colour. And therefore, if by refraction, or any other of the aforesaid causes, the difform Rays, latent in such a mixture, be separated, there shall emerge colours different from the colour of the composition. Which colours are not New generated, but only made Apparent by being parted; for if they be again intirely mix't and blended together, they will again compose that colour, which they did before separation. And for the same reason, Transmutations made by the convening of divers colours are not real; for when the difform Rays are again severed, they will exhibit the very same colours, which they did before they entered the composition; as you see, Blew and Yellow powders, when finely mixed, appear to the naked eye Green, and yet the colours of the Component corpuscles are not thereby really transmuted, but only blended. For, when viewed with a good Microscope, they still appear Blew and Yellow interspersedly.

5. There are therefore two sorts of colours. The one original and simple, the other compounded of these. The Original or primary colours are, Red, Yellow, Green, Blew, and a Violet-purple, together with Orange, Indico, and an indefinite variety of Intermediate gradations.

6. The same colours in Specie with these Primary ones may be also produced by composition: For, a mixture of Yellow and Blew makes Green; of Red and Yellow makes Orange; of Orange and Yellowish green makes yellow. And in general, if any two Colours be mixed, which in the series of those, generated by the Prisme, are not too far distant one from another, they by their mutual alloy compound that colour, which in the said series appeareth in the mid-way between them. But those, which are situated at too great a distance, do not so. Orange

and Indico produce not the intermediate Green, nor Scarlet and Green the intermediate yellow.

7. But the most surprising and wonderful composition was that of Whiteness. There is no one sort of Rays which alone can exhibit this. 'Tis ever compounded, and to its composition are requisite all the aforesaid primary Colours, mixed in a due proportion. I have often with Admiration beheld, that all the Colours of the Prisme being made to converge, and thereby to be again mixed as they were in the light before it was Incident upon the Prisme, reproduced light, intirely and perfectly white, and not at all sensibly differing from the direct Light of the Sun, unless when the glasses, I used, were not sufficiently clear; for then they would a little incline it to their colour.

8. Hence therefore it comes to pass, that Whiteness is the usual colour of Light; for, Light is a confused aggregate of Rays indued with all sorts of Colors, as they are promiscuously darted from the various parts of luminous bodies. And of such a confused aggregate, as I said, is generated Whiteness, if there be a due proportion of the Ingredients; but if any one predomi-nate, the Light must incline to that colour; as it happens in the Blew flame of Brimstone; the yellow flame of a Candle; and the various colours of the Fixed stars.

9. These things considered, the manner, how colours are produced by the Prisme, is evident. For, of the Rays, consti-tuting the incident light, since those which differ in Colour proportionally differ in refrangibility, they by their unequal refractions must be severed and dispersed into an oblong form in an orderly succession from the least refracted Scarlet to the most refracted Violet. And for the same reason it is, that objects, when looked upon through a Prisme, appear coloured. For, the difform Rays, by their unequal Refractions, are made

to diverge towards several parts of the Retina, and there express
the Images of things coloured, as in the former case they did
the Suns Image upon a wall. And by this inequality of refractions they become not only coloured, but also very confused
and indistinct.

10. Why the Colours of the Rainbow appear in falling
drops of Rain, is also from hence evident. For, those drops,
which refract the Rays, disposed to appear purple, in greatest quantity to the Spectators eye, refract the Rays of other
sorts so much less, as to make them pass beside it; and such
are the drops in the inside of the Primary Bow, and on the
outside of the Second or Exteriour one. So those drops, which
refract in greatest plenty the Rays, apt to appear red, toward the
Spectators eye, refract those of other sorts so much more, as to
make them pass beside it; and such are the drops on the exterior
part of the Primary, and interior part of the Secondary Bow.

11. The odd Phaenomena of an infusion of Lignum
Nephriticum, Leaf gold, Fragments of coloured glass, and
some other transparently coloured bodies, appearing in one
position of one colour, and of another in another, are on these
grounds no longer riddles. For, those are substances apt to
reflect one sort of light and transmit another; as may be seen
in a dark room, by illuminating them with similar or uncompounded light. For then they appear of that colour only, with
which they are illuminated, but yet in one position more vivid
and luminous than in another, accordingly as they are disposed
more or less to reflect or transmit the incident colour.

12. From hence also is manifest the reason of an unexpected
Experiment, which Mr. Hook somewhere in his Micrographia
relates to have made with two wedg-like transparent vessels,
fill'd the one with a red, the other with a blew liquor: namely,

that though they were severally transparent enough, yet both together became opake; For, if one transmitted only red, and the other only blew, no rays could pass through both.

13. I might add more instances of this nature, but I shall conclude with this general one, that the Colours of all natural Bodies have no other origin than this, that they are variously qualified to reflect one sort of light in greater plenty then another. And this I have experimented in a dark Room by illuminating those bodies with uncompounded light of divers colours. For by that means any body may be made to appear of any colour. They have there no appropriate colour, but ever appear of the colour of the light cast upon them, but yet with this difference, that they are most brisk and vivid in the light of their own day-light colour. Minium[a red lead pigment] appeareth there of any colour indifferently, with which 'tis illustrated, but yet most luminous in red, and so Bise[a light blue, copper-based pigment] appeareth indifferently of any colour with which 'tis illustrated, but yet most luminous in blew. And therefore Minium reflecteth Rays of any colour, but most copiously those indued with red; and consequently when illustrated with day-light, that is, with all sorts of Rays promiscuously blended, those qualified with red shall abound most in the reflected light, and by their prevalence cause it to appear of that colour. And for the same reason Bise, reflecting blew most copiously, shall appear blew by the excess of those Rays in its reflected light; and the like of other bodies. And that this is the intire and adequate cause of their colours, is manifest, because they have no power to change or alter the colour of any sort of Rays incident apart, but put on all colours indifferently, with which they are enlightened.

These things being so, it can be no longer disputed,

whether there be colours in the dark, nor whether they be the qualities of the objects we see, no nor perhaps, whether Light be a Body. For, since Colours are the qualities of Light, having its Rays for their intire and immediate subject, how can we think those Rays qualities also, unless one quality may be the subject of and sustain another; which in effect is to call it substance. We should not know Bodies for substances, were it not for their sensible qualities, and the Principal of those being now found due to something else, we have as good reason to believe that to be a substance also.

Besides, who ever thought any quality to be a heterogeneous aggregate, such as Light is discovered to be. But, to determine more absolutely, what Light is, after what manner refracted, and by what modes or actions it produceth in our minds the Phantasms of Colours, is not so easie. And I shall not mingle conjectures with certainties.

Reviewing what I have written, I see the discourse it self will lead to divers Experiments sufficient for its examination: And therefore I shall not trouble you further, than to describe one of those, which I have already insinuated.

In a darkened Room make a hole in the shut of a window, whose diameter may conveniently be about a third part of an inch, to admit a convenient quantity of the Sun's light: And there place a clear and colourless Prisme, to refract the entring light towards the further part of the Room, which, as I said, will thereby be diffused into an oblong coloured Image. Then place a Lens of about three foot radius (suppose a broad Object-glass of a three foot Telescope,) at the distance of about four or five foot from thence, through which all those colours may at once be transmitted, and made by its Refraction to convene at a further distance of about ten or twelve foot. If at that distance

you intercept this light with a sheet of white paper, you will
see the colours converted into whiteness again by being min-
gled. But it is requisite, that the Prisme and Lens be placed
steddy, and that the paper, on which the colours are cast, be
moved to and fro; for, by such motion, you will not only find,
at what distance the whiteness is most perfect, but also see, how
the colours gradually convene, and vanish into whiteness, and
afterwards having crossed one another in that place where they
compound Whiteness, are again dissipated, and severed, and
in an inverted order retain the same colours, which they had
before they entered the composition. You may also see, that, if
any of the Colours at the Lens be intercepted, the Whiteness
will be changed into the other colours. And therefore, that the
composition of whiteness be perfect, care must be taken, that
none of the colours fall beside the Lens.

[...]

If you proceed further to try the impossibility of changing
any uncompounded colour (which I have asserted in the third
and thirteenth Propositions,) 'tis requisite that the Room be
made very dark, least any scattering light, mixing with the colour, disturb and allay it, and render it compound, contrary to
the design of the Experiment. 'Tis also requisite, that there be
a perfecter separation of the Colours, than, after the manner
above described, can be made by the Refraction of one single
Prisme, and how to make further separations, will scarce be difficult to them, that consider the discovered laws of Refractions.
But if tryal shall be made with colours not throughly separated,
there must be allowed changes proportionable to the mixture.
Thus if compound Yellow light fall upon Blew Bise, the Bise
will not appear perfectly yellow, but rather green, because there
are in the yellow mixture many rays indued with green, and

Green being less remote from the usual blew colour of Bise than yellow, is the more copiously reflected by it.

In like manner, if any one of the Prismatick colours, suppose Red, be intercepted, on design to try the asserted impossibility of reproducing that Colour out of the others, which are pretermitted; 'tis necessary, either that the colours be very well parted before the red be intercepted, or that together with the red the neighbouring colours, into which any red is secretly dispersed, (that is, the yellow, and perhaps green too) be intercepted, or else, that allowance be made for the emerging of so much red out of the yellow & green, as may possibly have been diffused, and scatteringly blended in those colours. And if these things be observed, the new Production of Red, or any intercepted colour will be found impossible.

This, I conceive, is enough for an Introduction to Experiments of this kind; which if any of the R. Society shall be so curious as to prosecute, I should be very glad to be informed with what success: That, if any thing seem to be defective, or to thwart this relation, I may have an opportunity of giving further direction about it, or of acknowledging my errors, if I have committed any.

Your humble Servt

Isaac NEWTON

Leonard Euler
From *Letters to a German Princess*

This is the eighteenth of Euler's many letters to the
Princess d'Anhalt Dessau, written in response to her
request to introduce her to the world of science. Penned
on 10 June 1760, it disputes Newton's ideas on how light
works. The letters were translated into English by Henry
Hunter in 1795 and immediately became a bestseller.

However strange the doctrine of the celebrated Newton
may appear, that rays proceed from the sun, by a continual
emanation, it has, however, been so generally received, that it
requires an effort of courage to call it in question. What has
chiefly contributed to this, is, no doubt, the high reputation of
the great English philosopher, who first discovered the true
laws of the motion of the heavenly bodies: and it is this very
discovery which led him to the system of emanation.

Descartes, in order to support his theory, was under the
necessity of filling the whole space of the heavens with a subtle matter, through which the celestial bodies move at perfect
liberty. But it is well known that if a body moves in air, it must
meet with a certain degree of resistance; from which Newton
concluded that, however subtle the matter of the heavens may
be supposed, the planets must encounter some resistance to
their motions. But, said he, this motion is not subject to any
resistance: the immense space of the heavens, therefore, contains no matter.

A perfect vacuum, then universally prevails. This is one of
the leading doctrines of the Newtonian philosophy, that the
immensity of the universe contains no matter, in the spaces not

occupied by the heavenly bodies. This being laid down, there is between the sun and us, or at least from the sun down to the atmosphere of the earth, an absolute vacuum. In truth, the farther we ascend, the more subtle we find the air to be; from whence it would apparently follow, that at length the air would be entirely lost. If the space between the sun and the earth be an absolute vacuum, it is impossible that the rays should reach us in the way of communication, as the sound of a bell is transmitted by means of the air. For if the air, intervening between the bell and our ear, were to be annihilated, we should absolutely hear nothing, let the bell be struck ever so violently.

Having established, then, a perfect vacuum between the heavenly bodies, there remains no other opinion to be adopted, but that of emanation: which obliged Newton to maintain, that the sun, and all other luminous bodies, emit rays, which are always particles, infinitely small, of their mass, darted from them with incredible force. It must be such to a very high degree, in order to impress on rays of light that inconceivable velocity with which they come from the sun to us, in the space of eight minutes.

But let us see whether this theory be consistent with Newton's leading doctrine, which requires an absolute vacuum in the heavens, that the planets may encounter no manner of resistance to their motions. You must conclude on a moment's reflection, that the space in which the heavenly bodies revolve, instead of remaining a vacuum, must be filled with the rays, not only of the sun, but likewise of all the other stars which are continually passing through it, from every quarter, and in all directions, with incredible rapidity. The heavenly bodies which traverse these spaces, instead of encountering a vacuum, will meet with the matter of luminous rays in a terrible agitation, which must disturb these bodies in their motions, much more than if it were in a state of rest.

Thus Newton, apprehensive lest a subtle matter, such as Descartes imagined, should disturb the motion of the planets, has recourse to a very strange expedient, and quite contradictory to his own intention, as, on his hypothesis, the planets must be exposed to a derangement infinitely more considerable. I have already submitted to you several other insuperable objections to the system of emanation; and we have now seen that the principal, indeed the only reason, which could induce Newton to adopt it, is so self-contradictory as wholly to overturn it. All these considerations united, leave us no room to hesitate about the rejection of this strange system of the emanation of light, however respectable the authority of the philosopher who invented it.

Newton was, without doubt, one of the greatest geniuses that ever existed. His profound knowledge, and his acute penetration into the most hidden mysteries of nature, will be a just object of admiration to the present, and to every future age. But the errors of this great man should serve to admonish us of the weakness of the human understanding, which, after having soared to the greatest possible heights, is in danger of plunging into manifest contradiction.

Michael Faraday
Thoughts on Ray-Vibrations

Faraday was asked by a friend to note down his unscripted lecture on light, given when Wheatstone was alleged to have left Faraday to lecture for him. This was the result.

To Richard Phillips, Esq.

Dear Sir,

At your request I will endeavour to convey to you a notion of that which I ventured to say at the close of the last Friday-evening Meeting, incidental to the account I gave of Wheatstone's electro-magnetic chronoscope; but from first to last understand that I merely threw out as matter for speculation, the vague impressions of my mind, for I gave nothing as the result of sufficient consideration, or as the settled conviction, or even probable conclusion at which I had arrived.

The point intended to be set forth for consideration of the hearers was, whether it was not possible that vibrations which in a certain theory are assumed to account for radiation and radiant phaenomena may not occur in the lines of force which connect particles, and consequently masses of matter together; a notion which as far as is admitted, will dispense with the aether, which in another view, is supposed to be the medium in which these vibrations take place.

You are aware of the speculation which I some time since uttered respecting that view of the nature of matter which considers its ultimate atoms as centres of force, and not as so many little bodies surrounded by forces, the bodies being considered in the abstract as independent of the forces and capable of existing without them. In the latter view, these little particles have a definite form and a certain limited size; in the former view such is not the case, for that which represents size may be considered as extending to any distance to which the lines of force of the particle extend: the particle indeed is supposed to exist only by these forces, and where they are it is. The consideration of matter under this view gradually led me to look at the lines of force as being perhaps the seat of vibrations of radiant phenomena.

Another consideration bearing conjointly on the hypothetical view both of matter and radiation, arises from the comparison of the velocities with which the radiant action and certain powers of matter are transmitted. The velocity of light through space is about 190,000 miles in a second; the velocity of electricity is, by the experiments of Wheatstone, shown to be as great as this, if not greater: the light is supposed to be transmitted by vibrations through an aether which is, so to speak, destitute of gravitation, but infinite in elasticity; the electricity is transmitted through a small metallic wire, and is often viewed as transmitted by vibrations also. That the electric transference depends on the forces or powers of the matter of the wire can hardly be doubted, when we consider the different conductibility of the various metallic and other bodies; the means of affecting it by heat or cold; the way in which conducting bodies by combination enter into the constitution of non-conducting substances, and the contrary; and the actual existence of one elementary body, carbon, both in the conducting and non-conducting state. The power of electric conduction (being a transmission of force equal in velocity to that of light) appears to be tied up in and dependent upon the properties of the matter, and is, as it were, existent in them.

I suppose we may compare together the matter of the aether and ordinary matter (as, for instance, the copper of the wire through which the electricity is conducted), and consider them as alike in their essential constitution; i.e. either as both composed of little nuclei, considered in the abstract as matter, and of force or power associated with these nuclei, or else both consisting of mere centres of force, according to Boscovich's theory and the view put forth in my speculation; for there is no reason to assume that the nuclei are more requisite in the one case than in the other. It is true that the copper gravitates and the aether does not, and that therefore the copper is ponderable and the aether is not; but that cannot indicate the presence of nuclei in the copper more than in the aether, for of all the powers of matter gravitation is the one in which the force extends to the greatest possible distance from the supposed nucleus, being infinite in relation to the size of the latter, and reducing the nucleus to a mere centre of force. The smallest atom of matter on the earth acts directly on the smallest atom of matter in the sun, though they are 95,000,000 miles apart; further, atoms which, to our knowledge, are at least nineteen times that distance, and indeed in cometary masses, far more, are in a similar way tied together by the lines of force extending from and belonging to each. What is there in the condition of the particles of the supposed aether, if there be even only one such particle between us and the sun, that can in subtility and extent compare to this?

Let us not be confused by the ponderability and gravitation of heavy matter, as if they proved the presence of the abstract nuclei; these are due not to the nuclei, but to the force super-added to them, if the nuclei exist at all; and, if the aether particles be without this force, which according to the assumption is the case, then they are more material, in the abstract sense, than the matter of this our globe; for matter, according to the assumption, being made up of nuclei and force, the aether particles have in this respect proportionately more of the nucleus and less of the force.

On the other hand, the infinite elasticity assumed as belonging to the particles of the aether is as striking and positive a force of it as gravity is of ponderable particles, and produces in its way effects as great; in witness whereof we have all the varieties of radiant agency as exhibited in luminous, caloric, and actinic phaenomena.

Perhaps I am in error in thinking the idea generally formed of the aether is that its nuclei are almost infinitely small, and that such force as it has, namely its elasticity, is almost infinitely intense. But if such be the received notion, what then is left in the aether but force or centres of force? As gravitation and solidity do not belong to it, perhaps many may admit this conclusion; but what are gravitation and solidity? certainly not the weight and contact of the abstract nuclei. The one is the consequence of an attractive force, which can act at distances as great as the mind of man can estimate or conceive; and the other is the consequence of a repulsive force, which forbids for ever the contact or touch of any two nuclei; so that these powers or properties should not in any degree lead those persons who conceive of the aether as a thing consisting of force only, to think any otherwise of ponderable matter, except that it has more and other forces associated with it than the aether has.

In experimental philosophy we can, by the phaenomena presented, recognize various kinds of lines of force; thus there are the lines of gravitating force, those of electro-static induction, those of magnetic action, and others partaking of a dynamic character might be perhaps included. The lines of electric and magnetic action are by many considered as exerted through space like the lines of gravitating force. For my own part, I incline to believe that when there are intervening particles of matter (being themselves only centres of force), they take part in carrying on the force through the line, but that when there are none, the line proceeds through space. Whatever the view adopted respecting them may be, we can, at all events, affect these lines of force in a manner which may be conceived as partaking of the nature of a shake or lateral vibration. For suppose two bodies, A B, distant from each other and under mutual action, and therefore connected by lines of force, and let us fix our attention upon one resultant of force, having an invariable direction as regards space; if one of the bodies move in the least degree right or left, or if its power be shifted for a moment within the mass (neither of these cases being difficult to realise if A and B be either electric or magnetic bodies), then an effect equivalent to a lateral disturbance will take place in the resultant upon which we are fixing our attention; for, either it will increase in force whilst the neighbouring results are diminishing, or it will fall in force as they are increasing.

It may be asked, what lines of force are there in nature which are fitted to convey such an action and supply for the vibrating theory the place of the aether? I do not pretend to answer this question with any confidence; all I can say is, that I do not perceive in any part of space, whether (to use the common phrase) vacant or filled with matter, anything but forces and the lines in which they are exerted. The lines of weight or gravitating force are, certainly, extensive enough to answer in this respect any demand made upon them by radiant phae-nomena; and so, probably, are the lines of magnetic force: and then who can forget that Mossotti has shown that gravitation, aggregation, electric force, and electro-chemical action may all have one common connection or origin; and so, in their actions at a distance, may have in common that infinite scope which some of these actions are known to possess?

The view which I am so bold to put forth considers, therefore, radiation as a kind of species of vibration in the lines of force which are known to connect particles and also masses of matter together. It endeavours to dismiss the aether, but not the vibration. The kind of vibration which, I believe, can alone account for the wonderful, varied, and beautiful phaenomena of polarization, is not the same as that which occurs on the surface of disturbed water, or the waves of sound in gases or liquids, for the vibrations in these cases are direct, or to and from the centre of action, whereas the former are lateral. It seems to me, that the resultant of two or more lines of force is in an apt condition for that action which may be considered as equivalent to a lateral vibration; whereas a uniform medium, like the aether, does not appear apt, or more apt than air or water.

The occurrence of a change at one end of a line of force easily suggests a consequent change at the other. The propagation of light, and therefore probably of all radiant action, occupies time; and, that a vibration of the line of force should account for the phaenomena of radiation, it is necessary that such vibration should occupy time also. I am not aware whether there are any data by which it has been, or could be ascertained whether such a power as gravitation acts without occupying time, or whether lines of force being already in existence, such a lateral disturbance at one end as I have suggested above, would require time, or must of necessity be felt instantly at the other end.

As to that condition of the lines of force which represents the assumed high elasticity of the aether, it cannot in
this respect be deficient: the question here seems rather to
be, whether the lines are sluggish enough in their action to
render them equivalent to the aether in respect of the time
known experimentally to be occupied in the transmission of
radiant force.

The aether is assumed as pervading all bodies as well as
space: in the view now set forth, it is the forces of the atomic
centres which pervade (and make) all bodies, and also penetrate all space. As regards space, the difference is, that the
aether presents successive parts of centres of action, and the
present supposition only lines of action; as regards matter, the
difference is, that the aether lies between the particles and so
carries on the vibrations, whilst as respects the supposition, it
is by the lines of force between the centres of the particles that
the vibration is continued. As to the difference in intensity of
action within matter under the two views, I suppose it will be
very difficult to draw any conclusion, for when we take the
simplest state of common matter and that which most nearly
causes it to approximate to the condition of the aether, namely
the state of the rare gas, how soon do we find in its elasticity
and the mutual repulsion of its particles, a departure from the
law, that the action is inversely as the square of the distance!

And now, my dear Phillips, I must conclude. I do not think
I should have allowed these notions to have escaped from me,
had I not been led unawares, and without previous consideration, by the circumstances of the evening on which I had to
appear suddenly and occupy the place of another. Now that
I have put them on paper, I feel that I ought to have kept
them much longer for study, consideration, and, perhaps final

rejection; and it is only because they are sure to go abroad in one way or another, in consequence of their utterance on that evening, that I give shape, if shape it may be called, in this reply to your inquiry. One thing is certain, that any hypothetical view of radiation which is likely to be received or retained as satisfactory, must not much longer comprehend alone certain phaenomena of light, but must include those of heat and of actinic influence also, and even the conjoined phaenomena of sensible heat and chemical power produced by them. In this respect, a view, which is in some degree founded upon the ordinary forces of matter, may perhaps find a little consider-ation amongst the other views that will probably arise. I think it likely that I have made many mistakes in the preceding pages, for even to myself, my ideas on this point appear only as the shadow of a speculation, or as one of those impressions on the mind which are allowable for a time as guides to thought and research. He who labours in experimental inquiries knows how numerous these are, and how often their apparent fitness and beauty vanish before the progress and development of real natural truth.

I am, my dear Phillips,

Ever truly yours,

M. Faraday,

April 15, 1846

John Tyndall
Faraday as a Discoverer

*Writing shortly after Faraday's death, John
Tyndall (himself an eminent scientist and Faraday's
successor at the Royal Institution) remembers the
beginnings of the great man. This contemporary
description is particularly helpful in understanding
Faraday the man. Why so much interest in
Faraday? Because he was in a pivotal position in
the history of our understanding with light and
Tyndall's chatty memoir gives a particularly good
feel for the way science was treated 150 years ago.*

It has been thought desirable to give you and the world
some image of Michael Faraday, as a scientific investigator and
discoverer. The attempt to respond to this desire has been to
me a labour of difficulty, if also a labour of love. For however
well acquainted I may be with the researches and discoveries of that great master —however numerous the illustrations
which occur to me of the loftiness of Faraday's character and
the beauty of his life —still to grasp him and his researches as
a whole; to seize upon the ideas which guided him, and connected them; to gain entrance into that strong and active brain,
and read from it the riddle of the world — this is a work not
easy of performance, and all but impossible amid the distrac-
tion of duties of another kind. That I should at one period or
another speak to you regarding Faraday and his work is natural,
if not inevitable; but I did not expect to be called upon to speak
so soon. Still the bare suggestion that this is the fit and proper
time for speech sent me immediately to my task: from it I have

returned with such results as I could gather, and also with the wish that those results were more worthy than they are of the greatness of my theme.

It is not my intention to lay before you a life of Faraday in the ordinary acceptation of the term. The duty I have to perform is to give you some notion of what he has done in the world; dwelling incidentally on the spirit in which his work was executed, and introducing such personal traits as may be necessary to the completion of your picture of the philosopher, though by no means adequate to give you a complete idea of the man.

The newspapers have already informed you that Michael Faraday was born at Newington Butts, on September 22, 1791, and that he died at Hampton Court, on August 25, 1867. Believing, as I do, in the general truth of the doctrine of hereditary transmission—sharing the opinion of Mr. Carlyle, that 'a really able man never proceeded from entirely stupid parents'—I once used the privilege of my intimacy with Mr. Faraday to ask him whether his parents showed any signs of unusual ability. He could remember none. His father, I believe, was a great sufferer during the latter years of his life, and this might have masked whatever intellectual power he possessed. When thirteen years old, that is to say in 1804, Faraday was apprenticed to a bookseller and bookbinder in Blandford Street, Manchester Square: here he spent eight years of his life, after which he worked as a journeyman elsewhere.

You have also heard the account of Faraday's first contact with the Royal Institution; that he was introduced by one of the members to Sir Humphry Davy's last lectures, that he took notes of those lectures; wrote them fairly out, and sent them to Davy, entreating him at the same time to enable him to quit trade, which he detested, and to pursue science, which he loved. Davy was helpful to the young man, and this should never be forgotten: he at once wrote to Faraday, and afterwards, when an opportunity occurred, made him his assistant. Mr. Gassiot has lately favoured me with the following reminiscence of this time:

Clapham Common, Surrey
November 28, 1867.

My Dear Tyndall,

Sir H. Davy was accustomed to call on the late Mr. Pepys, in the Poultry, on his way to the London Institution, of which Pepys was one of the original managers; the latter told me that on one occasion Sir H. Davy, showing him a letter, said:

'Pepys, what am I to do, here is a letter from a young man named Faraday; he has been attending my lectures, and wants me to give him employment at the Royal Institution—what can I do?' 'Do?' replied Pepys, 'put him to wash bottles; if he is good for anything he will do it directly, if he refuses he is good for nothing.' 'No, no,' replied Davy; 'we must try him with something better than that.' The result was, that Davy engaged him to assist in the Laboratory at weekly wages.

Davy held the joint office of Professor of Chemistry and Director of the Laboratory; he ultimately gave up the former to the late Professor Brande, but he insisted that Faraday should be appointed Director of the Laboratory, and, as Faraday told me, this enabled him on subsequent occasions to hold a definite position in the Institution, in which he was always supported by Davy. I believe he held that office to the last.

Believe me, my dear Tyndall, yours truly,

J. P. Gassiot.

From a letter written by Faraday himself soon after his appointment as Davy's assistant, I extract the following account of his introduction to the Royal Institution:

London, Sept. 13, 1813.

As for myself, I am absent (from home) nearly day and night, except occasional calls, and it is likely shall shortly be absent entirely, but this (having nothing more to say, and at the request of my mother) I will explain to you. I was formerly a bookseller and binder, but am now turned philosopher, which happened thus:

Whilst an apprentice, I, for amusement, learnt a little chemistry and other parts of philosophy, and felt an eager desire to proceed in that way further. After being a journeyman for six months, under a disagreeable master, I gave up my business, and through the interest of a Sir H. Davy, filled the situation of chemical assistant to the Royal Institution of Great Britain, in which office I now remain; and where I am constantly employed in observing the works of nature, and tracing the manner in which she directs the order and arrangement of the world. I have lately had proposals made to me by Sir Humphry Davy to accompany him in his travels through Europe and Asia, as philosophical assistant. If I go at all I expect it will be in October next—about the end; and my absence from home will perhaps be as long as three years. But as yet all is uncertain.

This account is supplemented by the following letter, written by Faraday to his friend De la Rive, on the occasion of the death of Mrs. Marcet. The letter is dated September 2, 1858:

My Dear Friend,

Your subject interested me deeply every way; for Mrs. Marcet was a good friend to me, as she must have been to many of the human race. I entered the shop of a bookseller and bookbinder at the age of thirteen, in the year 1804, remained there eight years, and during the chief part of my time bound books. Now it was in those books, in the hours after work, that I found the beginning of my philosophy.

There were two that especially helped me, the *Encyclopaedia Britannica*, from which I gained my first notions of electricity, and Mrs. Marcet's *Conversation on Chemistry*, which gave me my foundation in that science.

Do not suppose that I was a very deep thinker, or was marked as a precocious person. I was a very lively imaginative person, and could believe in the *Arabian Nights* as easily as in the Encyclopaedia. But facts were important to me, and saved me. I could trust a fact, and always cross-examined an assertion. So when I questioned Mrs. Marcet's book by such little experiments as I could find means to perform, and found it true to the facts as I could understand them, I felt that I had got hold of an anchor in chemical knowledge, and clung fast to it. Thence my deep veneration for Mrs. Marcet—first as one who had conferred great personal good and pleasure on me; and then as one able to convey the truth and principle of those boundless fields of knowledge which concern natural things to the young, untaught, and inquiring mind.

You may imagine my delight when I came to know Mrs. Marcet personally; how often I cast my thoughts backward, delighting to connect the past and the present; how often, when sending a paper to her as a thank-offering, I thought of my first instructress, and such like thoughts will remain with me.

I have some such thoughts even as regards your own father; who was, I may say, the first who personally at Geneva, and afterwards by correspondence, encouraged, and by that sustained me.

Twelve or thirteen years ago Mr. Faraday and myself quitted the Institution one evening together, to pay a visit to our friend Grove in Baker Street. He took my arm at the door, and, pressing it to his side in his warm genial way, said, 'Come, Tyndall, I will now show you something that will interest you.' We walked northwards, passed the house of Mr. Babbage, which drew forth a reference to the famous evening parties once assembled there. We reached Blandford Street, and after a little looking about he paused before a stationer's shop, and then went in. On entering the shop, his usual animation seemed doubled; he looked rapidly at everything it contained. To the left on entering was a door, through which he looked down into a little room, with a window in front facing Blandford Street.

Drawing me towards him, he said eagerly, 'Look there, Tyndall, that was my working-place. I bound books in that little nook.' A respectable-looking woman stood behind the counter: his conversation with me was too low to be heard by her, and he now turned to the counter to buy some cards as an excuse for our being there. He asked the woman her name—her predecessor's name—his predecessor's name. 'That won't do,' he said, with good-humoured impatience; 'who was his predecessor?' 'Mr. Riebau,' she replied, and immediately added, as if suddenly recollecting herself, 'He, sir, was the master of Sir Charles Faraday.' 'Nonsense!' he responded, 'there is no such person.' Great was her delight when I told her the name of her visitor; but she assured me that as soon as she saw him running about the shop, she felt—though she did not know why—that it must be 'Sir Charles Faraday.'

William Thomson, Baron Kelvin of Largs
On the Ether

In a lecture given in 1884, Thomson describes the ether in its final evolution before Michelson finally disproved its existence. The term 'luminiferous' merely means 'light carrying'.

I move through this 'luminiferous ether' as if it were nothing. But were there vibrations with such frequency in a medium of steel or brass, they would be measured by millions and millions and millions of tons' action on a square inch of matter. There are no such forces in our air. Comets make a disturbance in the air, and perhaps the luminiferous ether is split up by the motion of a comet through it. So when we explain the nature of electricity, we explain it by a motion of the luminiferous ether. We cannot say that it is electricity. What can this luminiferous ether be? It is something that the planets move through with the greatest ease. It permeates our air; it is nearly in the same condition, so far as our means of judging are concerned, in our air and in the interplanetary space.

The air disturbs it but little; you may reduce air by air – pumps to the hundred thousandth of its density, and you make little effect in the transmission of light through it. The luminiferous ether is an elastic solid, for which the nearest analogy I can give you is this jelly which you see, and the nearest analogy to the waves of light is the motion, which you can imagine, of this elastic jelly, with a ball of wood floating in the middle of it. Look there, when with my hand I vibrate the little red ball up and down, or when I turn it quickly round the vertical diameter, alternately in opposite directions – that is the nearest representation I can give you of the vibrations of luminiferous ether.

[At this point he showed a large bowl of clear jelly with a small red wooden ball embedded in the surface near the centre.]

Another illustration is Scottish shoemakers' wax or Burgundy pitch, but I know Scottish shoemakers' wax better. It is heavier than water, and absolutely answers my purpose. I take a large slab of the wax, place it in a glass jar filled with water, place a number of corks on the lower side and bullets on the upper side. It is brittle like the Trinidad pitch or Burgundy pitch which I have in my hand – you can see how hard it is – but when left to itself it flows like a fluid. The shoemakers' wax breaks with a brittle fracture, but it is viscous and gradually yields.

What we know of the luminiferous ether is that it has the rigidity of a solid and gradually yields. Whether or not it is brittle and cracks we cannot yet tell, but I believe the discoveries in electricity and the motions of comets and the marvellous spurts of light from them, tend to show cracks in the luminiferous ether – show a correspondence between the electric flash and the aurora borealis and cracks in the luminiferous ether. Do not take this as an assertion, it is hardly more than a vague scientific dream: but you may regard the existence of the luminiferous ether as a reality of science; that is, we have an all-pervading medium, an elastic solid, with a great degree of rigidity – a rigidity so prodigious in proportion to its density that the vibrations of light in it have the frequencies I have mentioned, with the wave-lengths I have mentioned. The fundamental question as to whether or not luminiferous ether has gravity has not been answered. We have no knowledge that the luminiferous ether is attracted by gravity; it is sometimes called imponderable because some people vainly imagine that it has no weight; I call it matter with the same kind of rigidity that this elastic jelly has.

Wilhelm Conrad Röntgen
ON A NEW KIND OF RAYS

A section of the paper given by Röntgen to the
Würzburg Physical and Medical Society in
1895, describing the discovery of X-rays.

A discharge from a large induction coil is passed through a Hittorf's vacuum tube, or through a well-exhausted Crookes' or Lenard's tube. The tube is surrounded by a fairly close-fitting shield of black paper; it is then possible to see, in a completely darkened room, that paper covered on one side with barium platinocyanide lights up with brilliant fluorescence when brought into the neighbourhood of the tube, whether the painted side or the other be turned towards the tube. The fluorescence is still visible at two metres distance. It is easy to show that the origin of the fluorescence lies within the vacuum tube.

It is seen, therefore, that some agent is capable of penetrating black cardboard which is quite opaque to ultra-violet light, sunlight, or arc-light. It is therefore of interest to investigate how far other bodies can be penetrated by the same agent. It is readily shown that all bodies possess this same transparency, but in very varying degrees. For example, paper is very transparent; the fluorescent screen will light up when placed behind a book of a thousand pages; printer's ink offers no marked resistance. Similarly the fluorescence shows behind two packs of cards; a single card does not visibly diminish the brilliancy of the light. So, again, a single thickness of tinfoil hardly casts a shadow on the screen; several have to be superposed to produce a marked effect. Thick blocks of wood are still transparent. Boards of pine two or three centimetres thick absorb only very little.

A piece of sheet aluminium, 15 mm. thick, still allowed the X-rays (as I will call the rays, for the sake of brevity) to pass, but greatly reduced the fluorescence. Glass plates of similar thickness behave similarly; lead glass is, however, much more opaque than glass free from lead. Ebonite several centimetres thick is transparent. If the hand be held before the fluorescent screen, the shadow shows the bones clearly with only faint outlines of the surrounding tissues.

Water and several other fluids are very transparent. Hydrogen is not markedly more permeable than air. Plates of copper, silver, lead, gold, and platinum also allow the rays to pass, but only when the metal is thin. Platinum .2 mm. thick allows some rays to pass; silver and copper are more transparent. Lead 1.5 mm. thick is practically opaque. If a square rod of wood 20 mm. in the side be painted on one face with white lead, it casts little shadow when it is so turned that the painted face is parallel to the X-rays, but a strong shadow if the rays have to pass through the painted side. The salts of the metals, either solid or in solution, behave generally as the metals themselves.

The preceding experiments lead to the conclusion that the density of the bodies is the property whose variation mainly affects their permeability. At least no other property seems so marked in this connection. But that density alone does not determine the transparency is shown by an experiment wherein plates of similar thickness of Iceland spar, glass, aluminium, and quartz were employed as screens. Then the Iceland spar showed itself much less transparent than the other bodies, though of approximately the same density. I have not remarked any strong fluorescence of Iceland spar compared with glass.

Increasing thickness increases the hindrance offered to the

rays by all bodies. A picture has been impressed on a photographic plate of a number of superposed layers of tinfoil, like
steps, presenting thus a regularly increasing thickness. This
is to be submitted to photometric processes when a suitable
instrument is available.

[...]

Of special interest in this connection is the fact that photographic dry plates are sensitive to the X-rays. It is thus possible
to exhibit the phenomena so as to exclude the danger of error.
I have thus confirmed many observations originally made by
eye observation with the fluorescent screen. Here the power
of X-rays to pass through wood or cardboard becomes useful.
The photographic plate can be exposed to the action without
removal of the shutter of the dark slide or other protecting
case, so that the experiment need not be conducted in darkness.
Manifestly, unexposed plates must not be left in their box near
the vacuum tube.

It seems now questionable whether the impression on
the plate is a direct effect of the X-rays, or a secondary result
induced by the fluorescence of the material of the plate. Films
can receive the impression as well as ordinary dry plates.

I have not been able to show experimentally that the
X-rays give rise to any caloric effects. These, however, may
be assumed, for the phenomena of fluorescence show that the
X-rays are capable of transformation. It is also certain that all
the X-rays falling on a body do not leave it as such.

The retina of the eye is quite insensitive to these rays: the
eye placed close to the apparatus sees nothing. It is clear from
the experiments that this is not due to want of permeability on
the part of the structures of the eye.

[...]

It is known that Lenard, in his investigations on cathode rays, has shown that they belong to the ether, and can pass through all bodies. Concerning the X-rays the same may be said.

In his latest work, Lenard has investigated the absorption coefficients of various bodies for the cathode rays, including air at atmospheric pressure, which gives 4.10, 3.40, 3.10 for 1 cm., according to the degree of exhaustion of the gas in discharge tube. To judge from the nature of the discharge, I have worked at about the same pressure, but occasionally at greater or smaller pressures. I find, using a Weber's photometer, that the intensity of the fluorescent light varies nearly as the inverse square of the distance between screen and discharge tube. This result is obtained from three very consistent sets of observations at distances of 100 and 200 mm. Hence air absorbs the X-rays much less than the cathode rays. This result is in complete agreement with the previously described result, that the fluorescence of the screen can still be observed at 2 metres from the vacuum tube. In general, other bodies behave like air; they are more transparent for the X-rays than for the cathode rays.

A further distinction, and a noteworthy one, results from the action of a magnet. I have not succeeded in observing any deviation of the X-rays even in very strong magnetic fields.

The deviation of cathode rays by the magnet is one of their peculiar characteristics; it has been observed by Hertz and Lenard, that several kinds of cathode rays exist which differ by their power of exciting phosphorescence, their susceptibility of absorption, and their deviation by the magnet; but a notable deviation has been observed in all cases which have yet been

investigated, and I think that such deviation affords a characteristic not to be set aside lightly.

[...]

If one asks, what then are these X-rays; since they are not
cathode rays, one might suppose, from their power of exciting
fluorescence and chemical action, them to be due to ultra-violet
light. In opposition to this view a weighty set of considerations
presents itself. If X-rays be indeed ultra-violet light, then that
light must posses the following properties.

(a) It is not refracted in passing from air into water,
carbon bisulphide, aluminium, rock-salt, glass or zinc.

(b) It is incapable of regular reflection at the surfaces
of the above bodies.

(c) It cannot be polarised by any ordinary polarising
media.

(d) The absorption by various bodies must depend
chiefly on their density.

That is to say, these ultra-violet rays must behave quite
differently from the visible, infra-red, and hitherto known
ultra-violet rays.

These things appear so unlikely that I have sought for
another hypothesis.

A kind of relationship between the new rays and light rays
appears to exist; at least the formation of shadows, fluorescence,
and the production of chemical action point in this direction.
Now it has been known for a long time, that besides the transverse vibrations which account for the phenomena of light, it
is possible that longitudinal vibrations should exist in the ether,
and, according to the view of some physicists, must exist. It is
granted that their existence has not yet been made clear, and

their properties are not experimentally demonstrated. Should not the new rays be ascribed to longitudinal waves in the ether?

I must confess that I have in the course of this research made myself more and more familiar with this thought, and venture to put the opinion forward, while I am quite conscious that the hypothesis advanced still requires a more solid foundation.

Further reading

A very subjective collection of books to expand on the natural history of light.

Blish, James – *Doctor Mirabilis*. The science fiction writer James Blish produced a tour-de-force in this historical novel of Roger Bacon's life. Impressively researched, it gives a wonderful picture of the complexities of thirteenth century academic life.

Calder, Nigel – *Einstein's Universe*. A short and effective introduction to relativity.

Campbell, Lewis and Garnett, William – *The Life of James Clerk Maxwell*. Maxwell's good friend Campbell does not produce an unbiased view of the great man's life, but his closeness to the subject gives an unparalleled opportunity to understand the single most important figure in our understanding of light.

Clegg, Brian – *The First Scientist*. A biography of Roger Bacon, giving a lot more detail on Bacon's remarkable work.

Clegg, Brian – *The God Effect*. A detailed description of the background to quantum entanglement and the remarkable applications that it makes possible.

Clegg, Brian – *The Man Who Stopped Time*. Scientific biography of Eadweard Muybridge, focussing on the science of his stop motion photography and his life.

Clegg, Brian – *The Quantum Age*. Exploration of quantum theory and how its applications influence our lives, including a good account of the development of the laser.

David, Rosalie – *Cult of the Sun.* Useful background on Sun worship in ancient Egypt.

Ditchburn, R. W. – *Light.* An excellent textbook exploring in great depth the physics of light. Very technical – not for the fainthearted.

Euler, Leonard – *Letters of Euler to a German Princess.* Translated into English in the late 1700s by Henry Hunter, these letters remain a fascinating snapshot of the way science was treated at the time. Particularly fascinating when Hunter disagrees with Euler and interposes his own comments.

Feynman, Richard – *QED – The Strange Theory of Light and Matter. A* semi-popular exploration of the amazing world of quantum electrodynamics by the greatest physicist since Einstein.

Feynman, Richard – *Surely You're Joking Mr Feynman! As* well as being a great physicist, Richard Feynman was a superb storyteller, and this collection of anecdotes about his life, told to fellow physicist Ralph Leighton, are a joy to read.

Gleick, James – *Genius.* A good attempt at capturing the essence of Richard Feynman; less successful when it tries to explore the nature of genius, but good on the man.

Gleick, James – *Isaac Newton.* Not as good as *The Last Sorcerer* (White; see below) in capturing Newton, the man, but very good insight into his physics and his work as Master of the Royal Mint.

Gribbin, John and Mary – *Richard Feynman, A Life in Science.* A valuable complementary picture of Feynman's life and work, putting it into context in twentieth-century physics.

Harman, P. M. – *The Natural Philosophy of James Clerk Maxwell.* A detailed book on Maxwell's theories and the way that they were developed. Sometimes hard going, but valuable.

Herbert, Nick – *Faster than Light*. An examination of superluminal loopholes in physics. Predates Chiao and Nimtz's superluminal experiments, but shows various alternatives, explores EPR and lays the foundation for the workings of the superluminal signals. Becomes a little obscure occasionally, but largely readable.

Isaacson, Walter – *Einstein: His Life and Universe*. Probably the best exploration of Einstein's life. Not too technical, but with enough science to cover his work too.

Moore, Patrick – *Eyes of the Universe*. A typically personal tour by leading amateur astronomer Patrick Moore through the history of telescopes from the very first through to the instruments of the late 1990s. Issued for the 40th anniversary of his *Sky at Night* TV programme.

Newton, Sir Isaac – *Opticks*. Available in a reprint dating back to the 1950s, Newton's classic is surprisingly readable, partly because it predates the rather precious way that modern scientific writing is always written passively ('it was observed that…').

Sabra, A. I. – *Theories of Light from Descartes to Newton*. The definitive exposition of one of the key periods of change in our understanding of light. Concentrates on the way Descartes, Huygens and Newton approached scientific discovery. Based on a PhD thesis, so a little dry.

Sobel, Michael I. – *Light*. Good general technical description of light and its workings.

White, Michael – *Isaac Newton, The Last Sorcerer*. A fascinating biography of Newton that digs below the legend that was already established by the end of Newton's life.

Williamson, Samuel J. and Cummins, Herman Z. – *Light and Color in Nature and Art.* An academic journey through the impact of light on nature and art. Describes well the impact of the science of light on both natural development and artistic interpretation.

Zajonc, Arthur – *Catching the Light.* Interesting exploration of the intersection of light and the mind.

Index